CHEMISTRY RESEARCH AND APPLICATIONS

BENZOTHIAZOLE

PREPARATION, STRUCTURE AND USES

CHEMISTRY RESEARCH AND APPLICATIONS

Additional books and e-books in this series can be found on Nova's website under the Series tab.

CHEMISTRY RESEARCH AND APPLICATIONS

BENZOTHIAZOLE

PREPARATION, STRUCTURE AND USES

ATAKAN HEIJSTEK
EDITOR

Copyright © 2020 by Nova Science Publishers, Inc.

All rights reserved. No part of this book may be reproduced, stored in a retrieval system or transmitted in any form or by any means: electronic, electrostatic, magnetic, tape, mechanical photocopying, recording or otherwise without the written permission of the Publisher.

We have partnered with Copyright Clearance Center to make it easy for you to obtain permissions to reuse content from this publication. Simply navigate to this publication's page on Nova's website and locate the "Get Permission" button below the title description. This button is linked directly to the title's permission page on copyright.com. Alternatively, you can visit copyright.com and search by title, ISBN, or ISSN.

For further questions about using the service on copyright.com, please contact:
Copyright Clearance Center
Phone: +1-(978) 750-8400   Fax: +1-(978) 750-4470   E-mail: info@copyright.com.

NOTICE TO THE READER

The Publisher has taken reasonable care in the preparation of this book, but makes no expressed or implied warranty of any kind and assumes no responsibility for any errors or omissions. No liability is assumed for incidental or consequential damages in connection with or arising out of information contained in this book. The Publisher shall not be liable for any special, consequential, or exemplary damages resulting, in whole or in part, from the readers' use of, or reliance upon, this material. Any parts of this book based on government reports are so indicated and copyright is claimed for those parts to the extent applicable to compilations of such works.

Independent verification should be sought for any data, advice or recommendations contained in this book. In addition, no responsibility is assumed by the Publisher for any injury and/or damage to persons or property arising from any methods, products, instructions, ideas or otherwise contained in this publication.

This publication is designed to provide accurate and authoritative information with regard to the subject matter covered herein. It is sold with the clear understanding that the Publisher is not engaged in rendering legal or any other professional services. If legal or any other expert assistance is required, the services of a competent person should be sought. FROM A DECLARATION OF PARTICIPANTS JOINTLY ADOPTED BY A COMMITTEE OF THE AMERICAN BAR ASSOCIATION AND A COMMITTEE OF PUBLISHERS.

Additional color graphics may be available in the e-book version of this book.

Library of Congress Cataloging-in-Publication Data

ISBN: 978-1-53617-548-6

Published by Nova Science Publishers, Inc. † New York

CONTENTS

Preface		vii
Chapter 1	Synthesis of Benzothiazoles *Syed Shafi and Ahmed Kamal*	1
Chapter 2	Benzathiazole Analogs as Anticonvulsant and Anticancer Agents *Ajayrajsinh R. Zala, Azazahemad Kureshi and Premlata Kumari*	71
Chapter 3	ESIPT Inspired Benzothiazole Fluorescent Molecules *Vikas Patil, Bhavana V. Mohite, Satish V. Patil and Sharad Patil*	99
Chapter 4	Benzothiazole: A Promising Scaffold for the Development of Anticancer Agents *Shah Alam Khan, Asif Husain, Yaseen Moosa Al Lawatia and Saif Ahmad*	117
Index		193

PREFACE

Benzothiazole is an aromatic heterocyclic compound with the chemical formula of C7H5NS. Benzothiazole: Preparation, Structure and Uses opens with a review on the synthesis of benzothiazoles for their anticancer, antimicrobial, antidiabetic, anticonvulsant, anti-inflammatory, anthelmintic, analgesic, antiviral and antitubercular activities.

Following this, considering the versatile nature of benzothiazole, various benzothiazole derivatives with anticonvulsant and anticancer activity are discussed in detail.

A report on the design strategies, detailed photo-physical properties and applications of benzothiazoles is provided in an effort to help researchers overcome existing challenges in designing novel excited state intramolecular proton transfer benzothiazole fluorescent molecules.

In closing, the authors present recent updates on the usefulness and structure-activity relationship of benzothiazole derivatives for the development of novel anticancer agents.

Chapter 1 - Benzothiazole is an aromatic heterocyclic compound with the chemical formula of C_7H_5NS. Structurally benzothiazoles consist of a 5-membered 1,3-thiazole ring fused to a benzene ring. Benzothiazoles are commonly prepared by treating 2-mercaptoaniline with acid chlorides/ aldehydes/esters/carboxylic acids. Benzothiazole is a privileged bicyclic ring system with multiple applications. Even though, benzothiazole itself is not widely used; derivatives of benzothiazoles have attracted wide attention due to their useful biological and pharmacological properties including anticancer, antimicrobial, antidiabetic, anticonvulsant, anti-inflammatory, anthelmintic, analgesic, antiviral and antitubercular activities. A large number of therapeutic agents have been synthesized based on benzothiazole nucleus. Riluzole and pramipexole are some of the drugs bearing benzothiazole scaffold which are used to treat amyotrophic lateral sclerosis and Parkinson's disease respectively.

Chapter 2 - Benzothiazole is an aromatic bicyclic heterocycle with 1, 3-thiazole fused with a benzene ring. On account of its extended π-delocalization, it binds to DNA molecules through π-π interaction and exhibits a wide assortment of dynamic destinations. Owing to the bioorganic and medicinal significance, the benzothiazole moiety would be possibly giving dynamic pharmacophores with diversified activity. Benzothiazole derivatives were found to show adequacy against some intense illnesses like cancer, neurodegeneration, neuropathic torment, infectious diseases, epilepsy, etc. As it contains electronegative atoms, it is achievable for different moieties to form conjugates. Riluzole, 2-Amino-6-trifluoro methoxy benzothiazole hydrochloride is an anticonvulsant drug that can also inhibit GABA (gamma-aminobutyric acid) uptake with an IC_{50} of 43 μM. Porters, 5-fluoro benzothiazole derivatives are an antitumor agent in phase I clinical trial that elicits selective activity against human-derived carcinomas of the breast, ovarian and renal origin. Considering the versatile nature of benzothiazole, various benzothiazole derivatives with anticonvulsant and anticancer activity will be discussed in this book chapter.

Chapter 3 - Excited state intramolecular proton transfer (ESIPT) based 2- substituted benzothiazole fluorescent moleculeshavegainedconsiderable attention in the pastfew yearsas ausefulmolecule in high-tech and classical application. It was due to its desirable unique photo-physical properties induced due to the proton transfer in an excited state. The photo-physical properties of these benzothiazole ESIPT derivatives make them an interesting moiety and were studied as a function of pH and viscosity. High fluorescence quantum efficiencies and photo-stability in the micro-environment give enhancement in fluorescence. The observations are bound to the movement of the molecules in the solvent as non-bonding interactions with the surrounding environment in solution. It was also relevant to find out the relative fluorescence quantum yields of these derivatives by using secondary reference standards such as anthracene and fluorescence. The development of ESIPT chromophore with high fluorescence quantum efficiencies and a long fluorescence lifetime in the solid state for benzothiazoles is always a challenging issue due to the unpredictable mechanism of fluorescence in the solid state. Here a report on design strategies, detailed photo-physical properties, and their applications will help in assisting researchers to overcome existing challenges in designing novel ESIPT benzothiazole fluorescent molecules for promising applications. Present chapter deals with the developments of fluorescent benzothiazole ESIPT molecules with a focus on fluorescence properties.

Chapter 4 - Benzothiazole is an example of bicyclic ring system in which benzene and thiazole, a five membered heterocyclic ring with two hetero atoms *viz.* nitrogen and sulfur, are fused together. This versatile biologically active scaffold has been found to be present in various natural products of therapeutic importance. Literature review indicates that benzothiazole nucleus has played an important role in drug design and drug discovery process of newer drugs. Indeed, it serves as a core nucleus in few clinically useful drugs such as viozan, riluzole, ethoxazolamide, etc. Several other broad spectrum molecules developed based on this promising scaffold are currently under different phases of clinical trials. In lieu of its diverse pharmacological activities, simple structure, ease of synthesis and substitution with different functionalities, medicinal chemists consider this skeleton as a privileged scaffold for the development of safe and efficacious chemotherapeutic agents. This review provides recent updates on the usefulness, structure activity relation ship and potential of benzothiazole derivatives for the development of novel anticancer agents.

In: Benzothiazole
Editor: Atakan Heijstek

ISBN: 978-1-53617-548-6
© 2020 Nova Science Publishers, Inc.

Chapter 1

SYNTHESIS OF BENZOTHIAZOLES

*Syed Shafi[1] and Ahmed Kamal[2],**

[1]Department of Chemistry, School of Chemical and Lifescience
[2]Department of Pharmaceutical Chemistry,
School of Pharmaceutical Education and Research.
Jamia Hamdard, New Delhi, India

ABSTRACT

Benzothiazole is an aromatic heterocyclic compound with the chemical formula of C_7H_5NS. Structurally benzothiazoles consist of a 5-membered 1,3-thiazole ring fused to a benzene ring. Benzothiazoles are commonly prepared by treating 2-mercaptoaniline with acid chlorides/ aldehydes/esters/carboxylic acids. Benzothiazole is a privileged bicyclic ring system with multiple applications. Even though, benzothiazole itself is not widely used; derivatives of benzothiazoles have attracted wide attention due to their useful biological and pharmacological properties including anticancer, antimicrobial, antidiabetic, anticonvulsant, anti-inflammatory, anthelmintic, analgesic, antiviral and antitubercular activities. A large number of therapeutic agents have been synthesized based on benzothiazole nucleus. Riluzole and pramipexole are some of the

---

* Corresponding Author's E-mail: ahmedkamal915@gmail.com; pvc@jamiahamdard.ac.in.

drugs bearing benzothiazole scaffold which are used to treat amyotrophic lateral sclerosis and Parkinson's disease respectively.

Keywords: benzothiazoles, riluzole, pramipexole, firefly luciferin, pharmaceutical applications, industrial applications.

1. INTRODUCTION

Benzothiazoles (BT) **1** have attained great attention due to their broad spectrum of biological and biophysical (nonlinear optical materials, molecular dyads, chemosensors) properties. Over the decades, benzothiazole and its derivatives have been utilized in a large number of industrial and consumer products including vulcanization accelerators, dyes, corrosion inhibitors, insecticides, fungicides, herbicides, algicides, food flavoring agents, nonlinear optics (NLO) and ultraviolet (UV) light stabilizers. Benzothiazole derivatives have wide range of applications in organic synthesis towards the synthesis of molecules with structural diversity. Benzothiazole alkaloids are relatively rare in nature, possibly because of complications involved in their biosynthesis. They are present in a range of marine or terrestrial natural compounds that possess useful biological activities. Benzothiazole itself was first isolated in 1967 from the volatiles of American cranberries Vaccinium macrocarpon (also called large cranberry and bearberry). Luciferin, thiazo-rifamycins, dercitin-kuanoniamine, 6-hydroxybenzothiazole-5-acetic acid and erythrazoles are some of the well known examples of naturally occurring benzothiazoles. This chapter describes the diverse applications of benzothiazoles along with the popular synthetic protocols for their synthesis and their influence on environment.

2. STRUCTURE

Structurally, it is a bicyclic ring system in which five membered 1,3-thiazole ring is fused with benzene ring through 4,5-positions. It is

Synthesis of Benzothiazoles

completely a planar molecule which fulfills the Huckel's rule of (4n+2) π electrons. It is colorless and slightly viscous liquid.

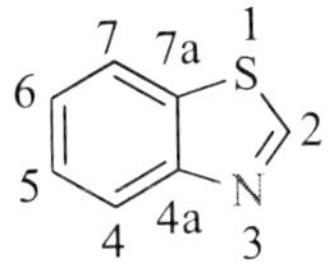

Figure 1. Core nucleus of benzothiazole.

2.1. Benzobisthiazoles

Molecules with benzfused bis-thiazoles have also been reported that have been formed by the fusion of the thiazole and the benzothiazole nuclei [1-4].

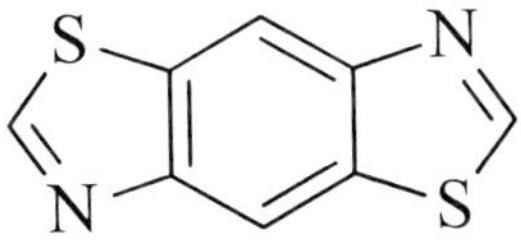

Figure 2. Benzobisthiazole.

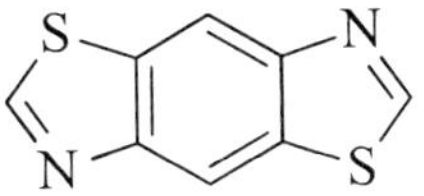

2
Benzo[1,2-d:4,5-d']bisthiazole

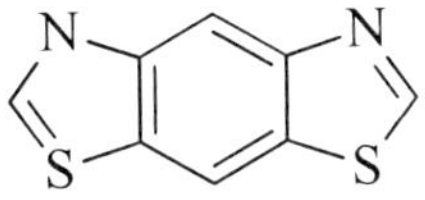

3
Benzo[1,2-d:5,4-d']bisthiazole

Figure 3. Structure of linear benzothiazoles.

On the basis of structure framework of benzothiazole, it is classified into two types (i) linearly fused benzothiazoles and (ii) angularly fused benzobisthiazoles [5]. Linear benzobisthiazoles include the benzo[1,2-*d*:4,5-*d'*]bisthiazole **2** (Type I) and benzo[1,2-*d*:5,4-*d'*]bisthiazole **3** (Type II) as shown in Figure 3.

While angular benzobisthiazoles include benzo[1,2-*d*:3,4-*d'*]bisthiazole **4** (Type III); benzo[1,2-*d*:4,3-*d'*]bisthiazole **5** (Type IV) and benzo[1,2-*d*:6,5-*d'*]bisthiazole **6** (Type V) as shown in Figure 4.

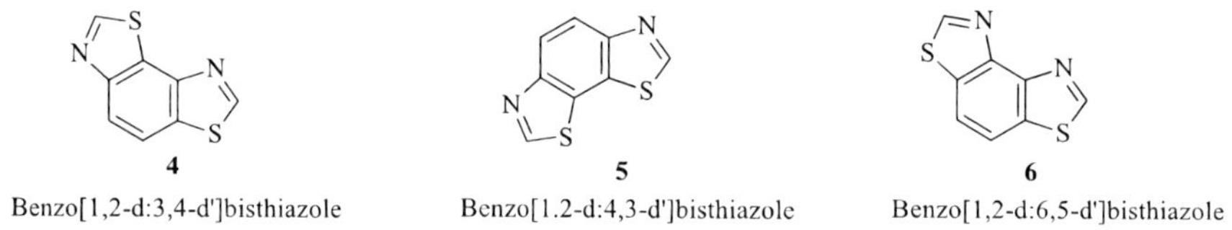

4
Benzo[1,2-d:3,4-d']bisthiazole

5
Benzo[1.2-d:4,3-d']bisthiazole

6
Benzo[1,2-d:6,5-d']bisthiazole

Figure 4. Structure of angular benzothiazoles.

2.2. Most Commonly Available Benzothiazoles

The most commonly available benzothiazoles are: 2-hydroxy-benzothiazole (2-OH-BTH) **7**, 2-amino-benzothiazole (2-NH$_2$-BTH) **8**, 2-methyl-benzothiazole (2-Me-BTH) **9**, 2-methylthio-benzothiazole (2-Me-S-BTH) **10**, 2-mercapto-benzothiazole (2-SH-BTH) **11**, 2-thiocyanomethylthio-benzothiazole (2-SCNMeS-BTH) **12**, and 2-benzothiazole-sulfonic acid (2-SO$_3$H-BTH) **13** as illustrated in Figure 5.

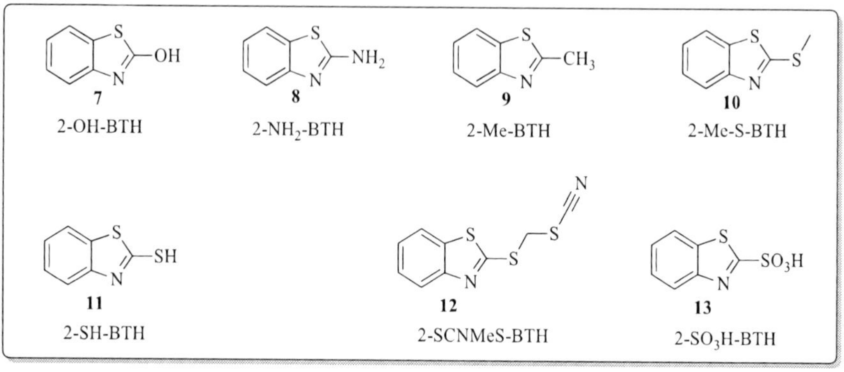

7
2-OH-BTH

8
2-NH$_2$-BTH

9
2-Me-BTH

10
2-Me-S-BTH

11
2-SH-BTH

12
2-SCNMeS-BTH

13
2-SO$_3$H-BTH

Figure 5. Most commonly available benzothiazoles.

3. NATURALLY OCCURRING BENZOTHIAZOLES

Benzothiazole alkaloids are relatively rare in nature, possibly because of complications involved in their biosynthesis. They are present in a range of marine or terrestrial natural compounds that have useful biological activities. Some of the naturally available benzothiazole derivatives have been depicted in Figure 6.

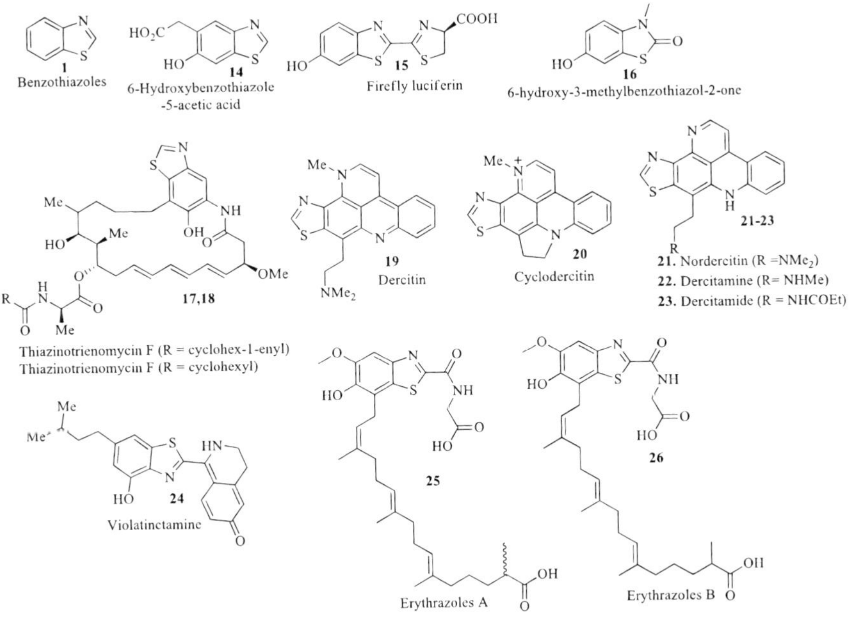

Figure 6. Naturally occurring benzothiazoles.

Benzothiazole itself was fist isolated by Anjou and Von Sydow in 1967 from the volatiles of American cranberries *Vaccinium macrocarpon* (also called large cranberry and bearberry) [6].

Since then it has been isolated from different sources including tail-gland of the red deer *Cervus elaphus* [7] (it was described as one of the sulfur volatiles in wines [8, 9]), volatile fraction of French oak wood (which is used in the aging of wine) [10], aroma fraction of tea leaves [11 and the fungi *Aspergillus clavatus* [12].

6-Hydroxybenzothiazole-5-acetic acid **14** is another naturally occurring benzothiazole derivative, which is isolated from the cultured filtrate of the bacterium *Actinosynnema sp.*[13] and *Paecilomyces lilacinus*[14]. It is also known as antibiotic C304A or M4582 and was found to be aldose reductase inhibitor. Firefly luciferin is a light-emitting biological pigment generally present in organs capable of bioluminescence. It was first isolated from the common American *firefly Photinus pyralis* in 1957 and a molecular formula of $C_{13}H_{12}N_2O_3S$ was assigned [15]. Later in 1963, White et al. have proposed another molecular formula for luciferin as $C_{11}H_8N_2O_3S_2$ and confirmed by its total synthesis [16].

Stierle et al. have reported the isolation of benzothiazole derivatives from the marine biosphere for the first time. They have isolated four relatively simple benzothiazole derivatives; 2-methylbenzothiazole **9**, 2-hydroxybenzothiazole **7**, 2-mercaptobenzothiazole **11** and 6-hydroxy-3-methylbenzothiazol-2-one **16** from fermentation culture extracts of *Micrococcus* sp., a marine bacterium obtained from tissues of the sponge *Tedaniaignis* [17].

Thiazinotrienomycin F (**17**) and thiazinotrienomycin G (**18**) are the two naturally occurring benzothiazoles isolated from the culture broth bacterium *Streptomycessp*. MJ672-m3 [18]. This natural compound belong to ansamycin family of antibiotics like the rifamycins and was found to exhibit anti-cancer activity [19].

Dercitin, cyclodercitin, nordercitin, dercitamine, dercitamide (Kuanoniamine C) (**19-23**), a pentacyclic structural benzothiazole based frameworks have been obtained from the *Dercitus sp*. Violatinctamine **24** is a bright orange compound which was isolated from a Kenyan marine tunicate of the genus *Cystodytes cf. violatinctus* in 2004.

Erythrazoles A (**25**) and B (**26**) are the anti-cancer agents which are extracted from an *Erythrobacter sp.* from mangrove sediments. They exhibit good cytotoxicity against a panel of non-small cell lung cancer cell lines H1325, H2122 and HCC366 with IC_{50} of 1.5, 2.5 and 6.8 µM respectively.

4. PREPARATION

Hofmann was the first one to report the synthesis of 2-substituted benzothiazole in 1887. Benzothiazoles have been synthesized by two major reactions i.e., Condensation and cyclization reactions as shown below (Figure 7) [20a].

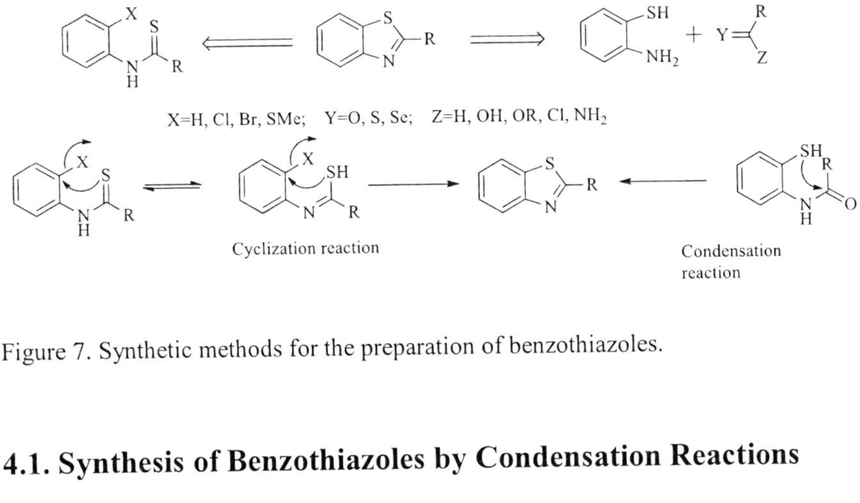

X=H, Cl, Br, SMe;   Y=O, S, Se;   Z=H, OH, OR, Cl, NH_2

Cyclization reaction

Condensation reaction

Figure 7. Synthetic methods for the preparation of benzothiazoles.

4.1. Synthesis of Benzothiazoles by Condensation Reactions

One of the traditional methods for preparation of BTA framework is the condensation of 2-aminobenzenethiol with substituted nitriles, aldehydes, carboxylic acids, acyl chlorides, or esters as illustrated in Scheme 1.

Different reagents including strong acids, Lewis acids, oxidative agents and different catalysts have been employed for this reaction [20b-d]. There are some special reagents that have been developed to catalyze this reaction like enzymatic reactions as given below.

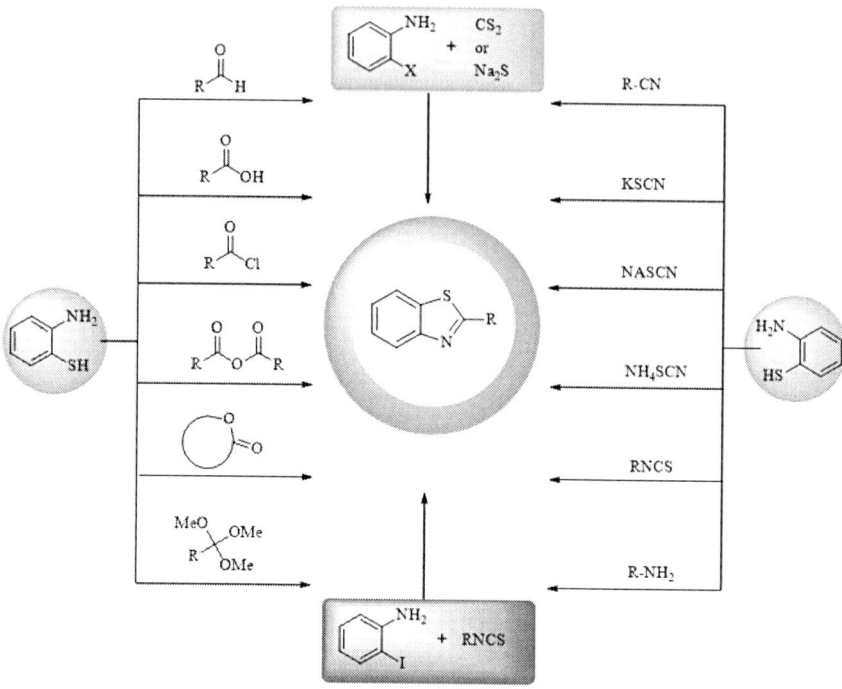

Scheme 1. General synthesis of benzothiazoles by condensation reactions.

4.1.1. Synthesis of Benzothiazoles Catalyzed by Baker's Yeast

This reaction involves the catalysis related to enzymes which causes condensation by forming either initial enzyme-2-aminothiophenol non-covalent complex or an enzyme-aldehyde complex.

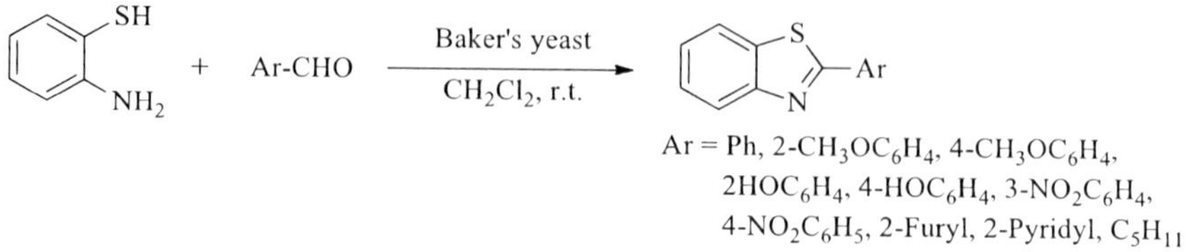

Ar = Ph, 2-CH$_3$OC$_6$H$_4$, 4-CH$_3$OC$_6$H$_4$, 2HOC$_6$H$_4$, 4-HOC$_6$H$_4$, 3-NO$_2$C$_6$H$_4$, 4-NO$_2$C$_6$H$_5$, 2-Furyl, 2-Pyridyl, C$_5$H$_{11}$

Scheme 2. Synthesis of benzothiazoles catalyzed by baker's yeast.

2-aminothiophenol and aldehydes is catalysed by bakers' yeast in dichloromethane solvent *via* forming either an initial enzyme-2-

aminothiophenol non-covalent complex or an enzyme aldehyde complex intermediate, this intermediate is treated with coenzymes such as FAD which promote the dehydrogenation and subsequent abstraction of a proton to yield 2-substituted benzothiazoles (given in Scheme 2) [21].

4.1.2. Synthesis of Benzothiazoles Using HTP/ZrP Catalysis

An efficient solvent free synthesis of 2-arylbenzothiazoles is also reported from the reaction of 2-aminothiophenol with various aldehydes in presence of tungstophosphoric acid using zirconium phosphate as catalyst in absence of solvent as illustrated in Scheme 3 [22].

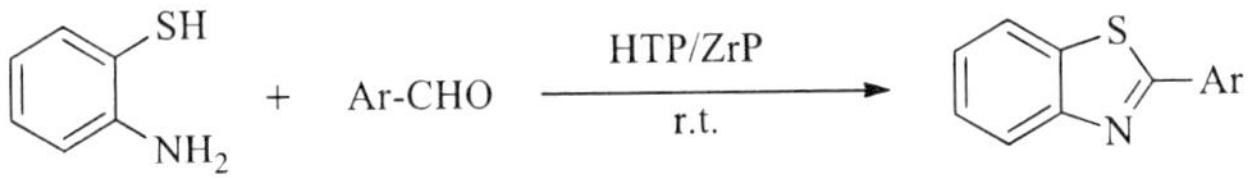

Ar = Ph, 2-CH$_3$OC$_6$H$_4$, 4-CH$_3$OC$_6$H$_4$, 2HOC$_6$H$_4$, 4-HOC$_6$H$_4$, 3-NO$_2$C$_6$H$_4$, 4-NO$_2$C$_6$H$_5$, 2-Furyl, 2-Pyridyl, C$_5$H$_{11}$

Scheme 3. Synthesis of benzothiazoles using HTP/ZrP catalyst.

4.1.3. Synthesis of 2-Substituted Benzothiazoles using H$_2$O$_2$/HCl System

Guo et al. have developed a simple method for the rapid synthesis of benzothiazoles from 2-aminothiophenol and arenealdehydes using H$_2$O$_2$/HCl system as illustrated in Scheme 4.

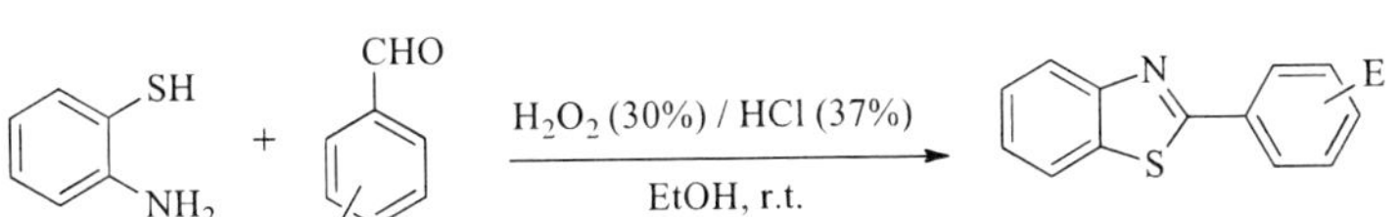

R = H, 2-OH, 4-OH, 3-NO$_2$, 4-NO$_2$, 3,4-di-OCH$_3$, 2,4-di-Cl, 3,5-di-C(CH$_3$)$_3$-4-OH, 1-naphthyl, 5-NO$_2$-1-naphthyl, 9-anthrayl

Scheme 4. Synthesis of 2-substituted benzothiazoles using H2O2/HCl system.

4.1.4. Enzymatic Synthesis of Benzothiazoles

Kumar et al. have reported the synthesis of 2-arylbenzothiazoles employing biocatalysis as demonstrated in Scheme 5 [24]. Arenealdehydes are reacted with 2-aminothiophenol in presence of glucose oxidase (GOX)-chloroperoxidase (CPO) catalytic enzymatic system under oxygen atmosphere. This is an efficient and eco-friendly method for the synthesis of 2-arylbenzothiazoles.

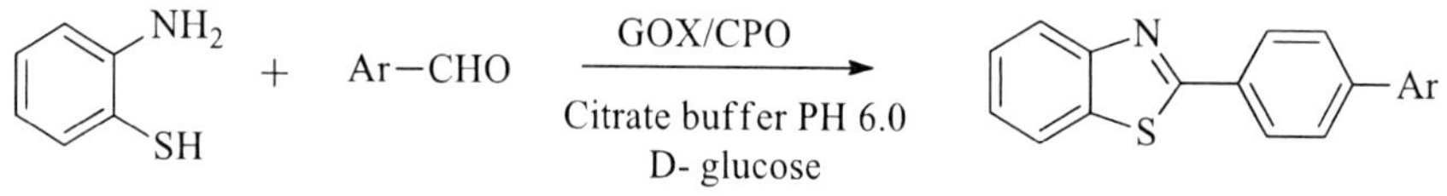

Scheme 5. Enzymatic synthesis of benzothiazoles.

4.1.5. Synthesis of Benzothiazoles Using RuCl₃ as Catalyst

Fan et al. have developed RuCl₃ catalyzed reactions for the synthesis of 2-substituted benzothiazoles as shown in Scheme 6. In this reactions 2-aminothiophenols underwent oxidative condensations with aldehydes to form 2-substituted benzothiazoles. This method involves the green solvent, the use of air as oxidant and the possibility of reusing RuCl₃/[bmim]PF6 (1-butyl-3-methylimidazolium hexafluorophosphate) to yield benzothiazoles in moderate to good yields. However, this method has a limitation due to oxidizable nature of 2-aminothiophenols bearing substituents.

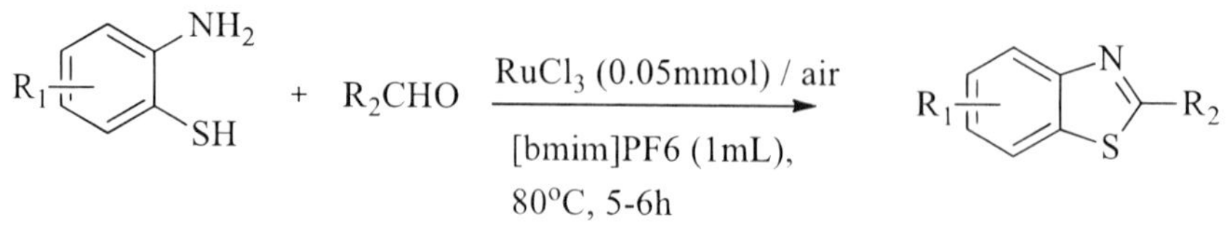

Scheme 6. Synthesis of benzothiazoles using RuCl3 as catalyst.

4.1.6. Synthesis of Benzothiazoles Using CTAB as Catalyst

Yang et al. have developed a green protocol for the synthesis of 2-substituted benzothiazoles in good to excellent yields [25]. This reaction involves the condensation of 2-aminothiophenol with different aldehydes using cetyltrimethyl ammonium bromide (CTAB) as a catalyst in water. This

Synthesis of Benzothiazoles

is one of the simplest methods, where water is used as a solvent and a cost-effective catalyst is used as well with simple workup.

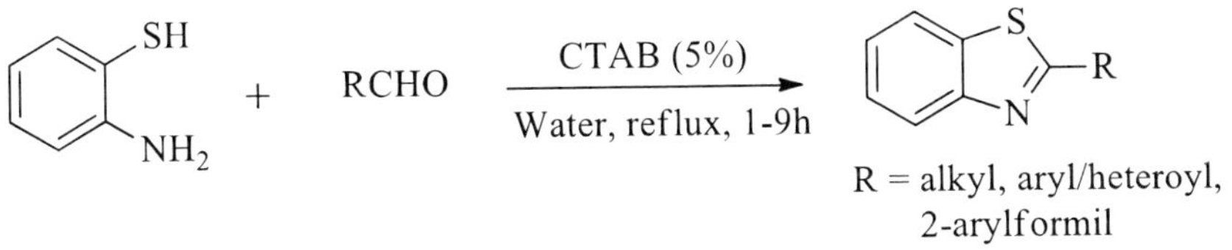

Scheme 7. Synthesis of benzothiazoles using CTAB as catalyst.

4.1.7. Combinatorial Synthesis of Benzothiazoles Using a Traceless Aniline Linker

Combinatorial synthesis of benzothiazoles using traceless aniline linker has been developed by Hioki et al. as shown in Scheme 8 [26].

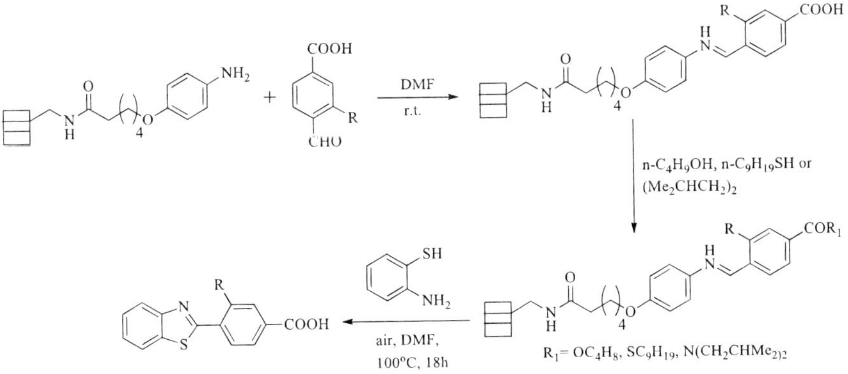

Scheme 8. Combinatorial synthesis of benzothiazoles.

4-Formylbenzoic acid when condensed with an aniline linker on the solid support forms resin-bound azomethine, which is further reacted with alcohols, thiols and amines to form azomethines. These azomethines are cleaved under neutral conditions by 2-aminothiophenols to yield benzothiazoles. This methodology involves solid-phase synthesis without the use of oxidants.

4.1.8. Solvent Free Synthesis

2-Aminothiophenol condensed with saturated olefinic fatty acids under solvent-free microwave irradiation conditions to form 2-substituted benzothiazoles in the presence of catalyst P_4S_{10}. This reaction takes about 3-4 min to afford high yield of the compound (as shown in Scheme 9).

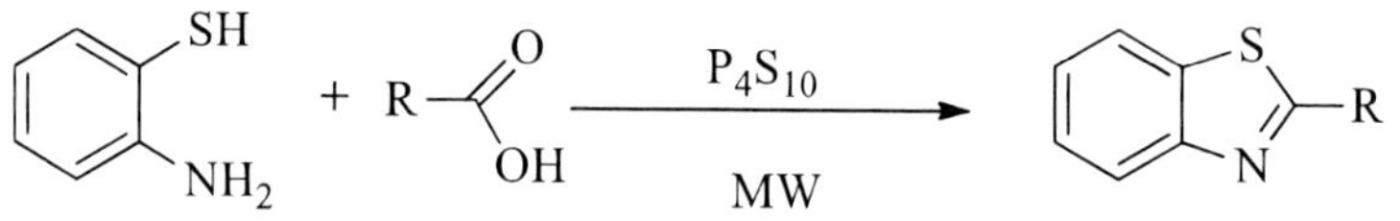

Scheme 9. Synthesis of 2-substituted benzothiazoles in solvent-free condition.

4.2. Synthesis of Benzothiazoles through Cyclization Reactions

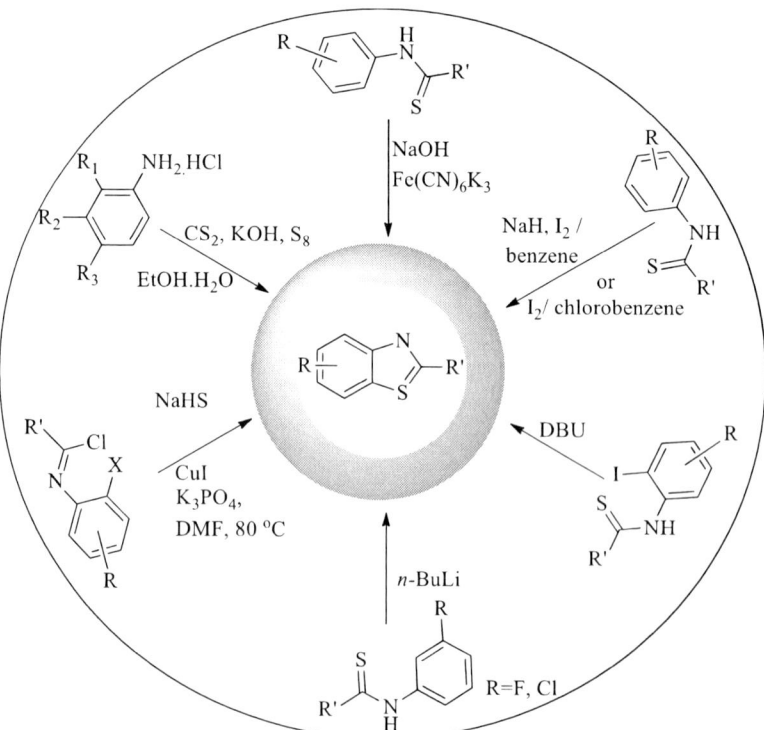

Scheme 10. Synthesis of benzothiazoles by cyclization reactions.

Synthesis of Benzothiazoles

Benzothiazoles have been synthesized through cyclization reactions of N-phenylthioamides or N-(2-halophenyl)thioamides as shown above.

4.2.1. Jacobson's Cyclization

Another method for preparing benzothiazoles is Jacobson's cyclization of thiobenzanilides (as shown in Scheme 11) [27]. It involves the conversion of thioanilides to benzothiazoles in presence of ferricyanide under basic conditions. This reaction can proceed through two mechanism, one is *via* thiol radical and the other is *via* thiomidic cation.

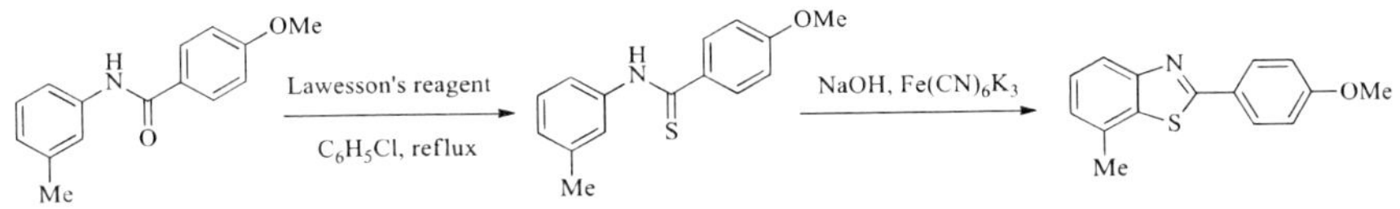

Scheme 11. Jacobson's cyclization.

4.2.2. Synthesis of Benzothiazoles Mediated by Iodine

Benzothiazoles has also been synthesized from thiobenzamides by the use of iodine without an *ortho* alkoxy or ester group as illustrated in scheme 12 [28]. In this method, thiobenzamides are treated with iodine in refluxing chlorobenzene or heating with sodium hydride and iodine in refluxing benzene for 2h leading to the formation of benzothiazoles in yields of 87%. While non-substituted thiobenzamides or bearing an electron-attractive group, such as bromine do not provide the expected benzothiazoles.

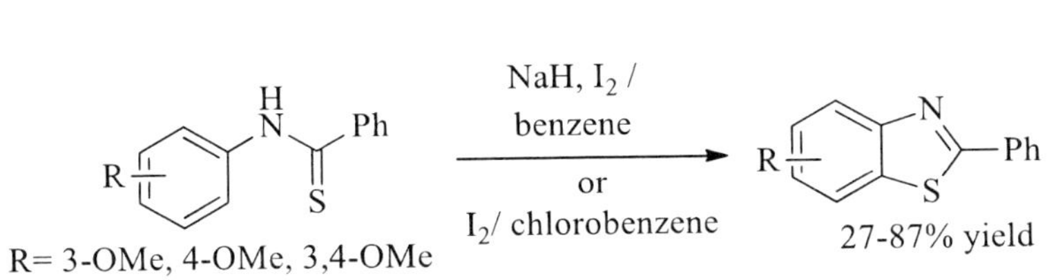

Scheme 12. Benzothiazoles synthesis mediated by iodine.

4.2.3. Pd-Catalyzed Synthesis of Benzothiazoles from 2-Bromoanilides

Itoh and Mase reported a new method for the synthesis of benzothiazoles from 2-bromoanilides with a thiol surrogate coupling reaction as demonstrated in Scheme 13 [29].

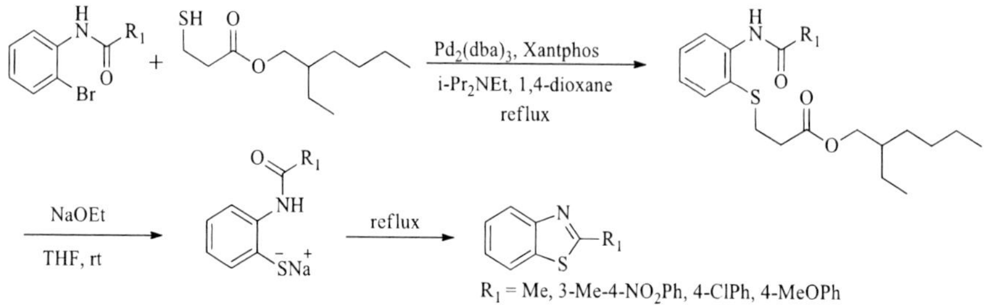

Scheme 13. Synthesis of benzothiazoles from 2-bromoanilides.

Firstly, 2-bromoanilides are reacted with 2-ethylhexyl 3-mercaptopropionate as a thiol surrogate in presence of $Pd_2(dba)_3$/Xantphos (Xantphos: a bidentate bis-phosphine ligand) to form a sulfite intermediate. Further on, this intermediate is reacted with NaOEt in THF solvent at room temperature to afford the corresponding sodium thiolate that undergoes intramolecular cyclization to afford desired benzothiazoles in good yields under refluxing conditions.

4.2.4. Synthesis of Benzothiazoles Using Lawesson's Reagent

Ding et al. have developed a one-pot synthetic strategy for the synthesis of benzothiazoles as illustrated in Scheme 14. In this process, 2-iodoanilines are treated with acid chlorides in the presence of Lawesson's reagent to afford benzothiazoles, more over when this was carried out in presence of DBU, it gave good yields.

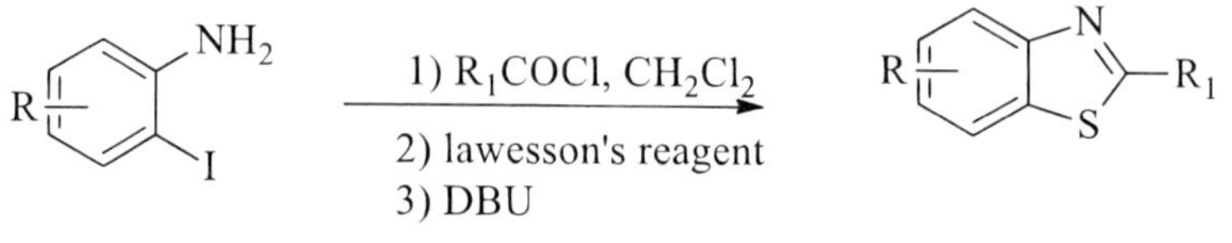

Scheme 14. Synthesis of benzothiazoles using Lawesson's reagent.

4.2.5. Synthesis of 2-Trifluoromethyl Benzothiazoles

Li et al. have prepared 2-substituted benzothiazoles by a one-pot synthesis through copper-catalyzed thiolation annulations of 1,4-dihalides with sodium hydrosulfide. 2,2,2-trifluoro-*N*-(2-haloaryl)acetimidoyl chlorides when reacted with NaHS in presence of CuI and K_3PO_4 afforded benzothiazoles in moderate to good yields as mentioned in Scheme 15 [30].

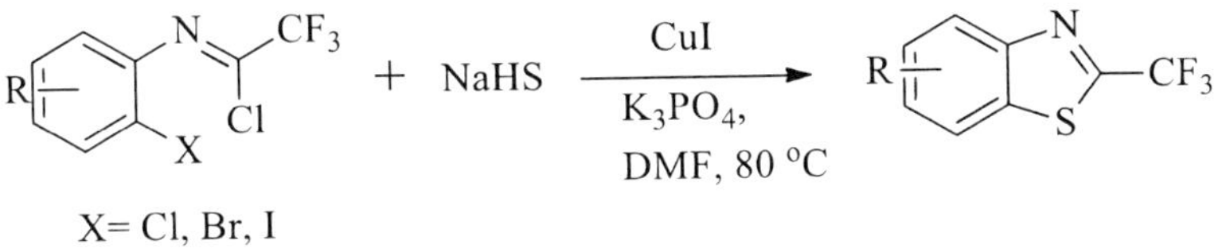

Scheme 15. Cu(I) catalysed synthesis of benzothiazole derivatives.

4.2.6. Synthesis of 2-Trifluoromethyl Benzothiazoles Using PdCl$_2$ as Catalyst

Zhu et al. also followed the above mentioned Scheme for synthesis of 2-trifluoromethylbenzothiazoles from trifluoroimidoyl chlorides and sodium hydrosulfide by employing $PdCl_2$ as a catalyst instead of CuI and no additives or oxidants are required for this reaction (Scheme 16) [31].

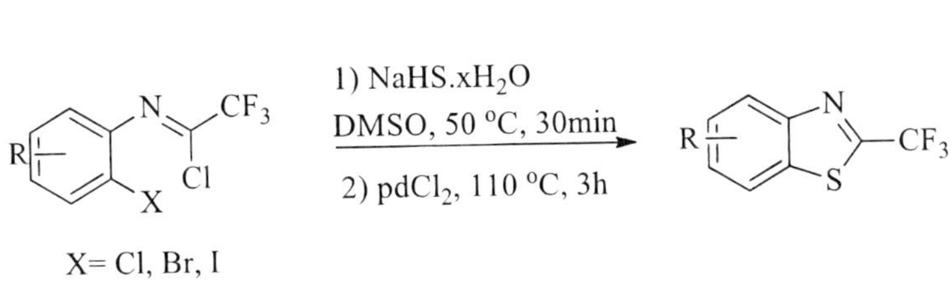

Scheme 16. PdCl$_2$ catalysed synthesis of benzothiazole derivatives.

4.2.7. Cyclization of Aniline Hydrochloride Using CS$_2$ and KOH

Synthesis of substituted 2-mercaptobenzothiazoles has been performed in two steps as illustrated in Scheme 17. In the first step, substituted anilines are converted to its hydrochloride salts and in the second step, this hydrochloride salt is reacted with carbon disulphide in the presence of sulphur in alkaline medium to yield 2-mercaptobenzothiazoles.

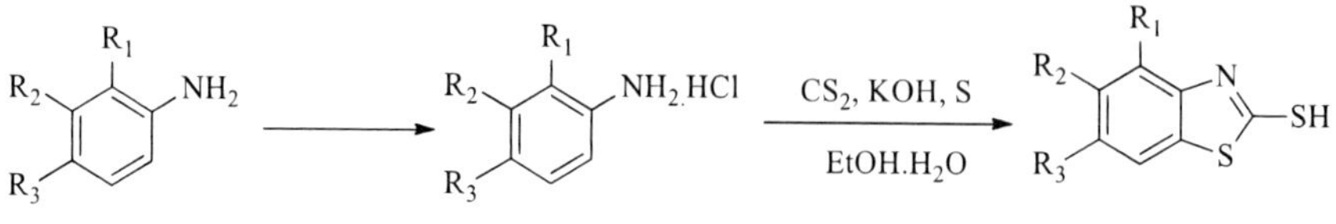

Scheme 17. Synthesis of substituted 2-mercapto benzothiazoles *via* cyclization.

5. APPLICATIONS OF BENZOTHIAZOLE DERIVATIVES

Benzothiazole is a privileged scaffold in drug discovery due to its broad spectrum of biological activities such as antidiabetic, anticancer, antimicrobial, anticonvulsant, anti-inflammatory, antiviral, antihelmintic, antileishmanial, schictosomicidal, antifungal, anti-psychoti and antitubercular properties [32].

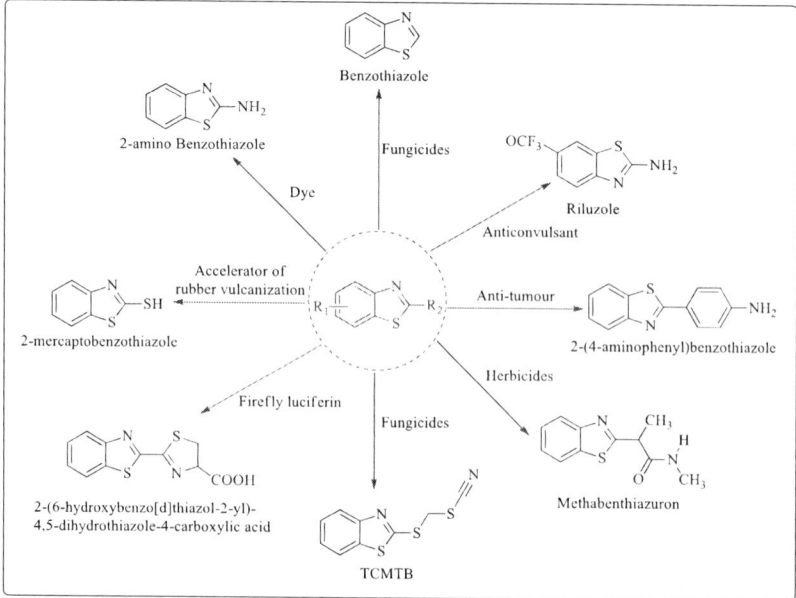

Figure 8. Pharmacological profile of benzothiazole.

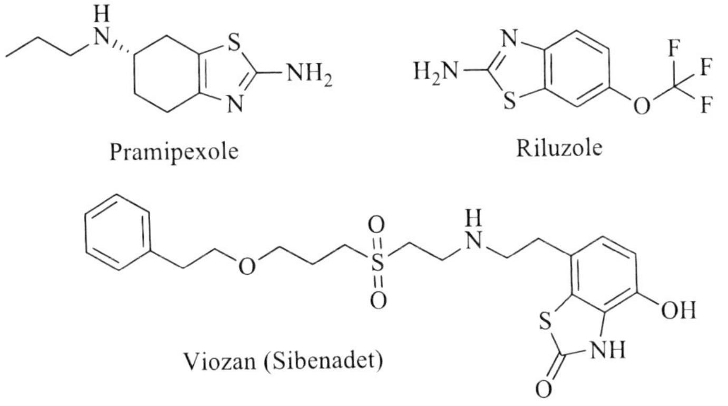

Figure 8a. Pharmacologically important benzothiazole derivatives.

Moreover, it also possesses remarkable chemical properties; it has been used as vulcanization accelerators, antioxidants and dopant in light emitting organic electroluminescent devices in industries. They are one of the important pharmacophores which comprises of potential applications in fields such as nonlinear optics (NLO), organic light emitting diodes, organic field-effect transistors, polymers, liquid crystals, dye-sensitized solar cells, fluorescent dyes, photonucleases, antioxidants and insecticides. Several benzothiazole derivatives are also used commercially as shown in Figure 8. Some of the marketed compounds like Tribunil and Ormet are used as herbicides and slimicides in paper and pulp industry. Riluzole (trade name: Rilutek) is a marketed drug and is used to treat amyotrophic lateral sclerosis. Pramipexole is used to treat Parkinson's disease. Viozan, a benzothiazole based drug candidate is under phase-III clinical trials for chronic obstructive pulmonary disease (COPD).

5.1. Pharmacological Applications of Benzothiazole Derivatives

Benzothiazole heterocycle are an important class of therapeutic agents that demonstrated a wide range of pharmacological activities [32].

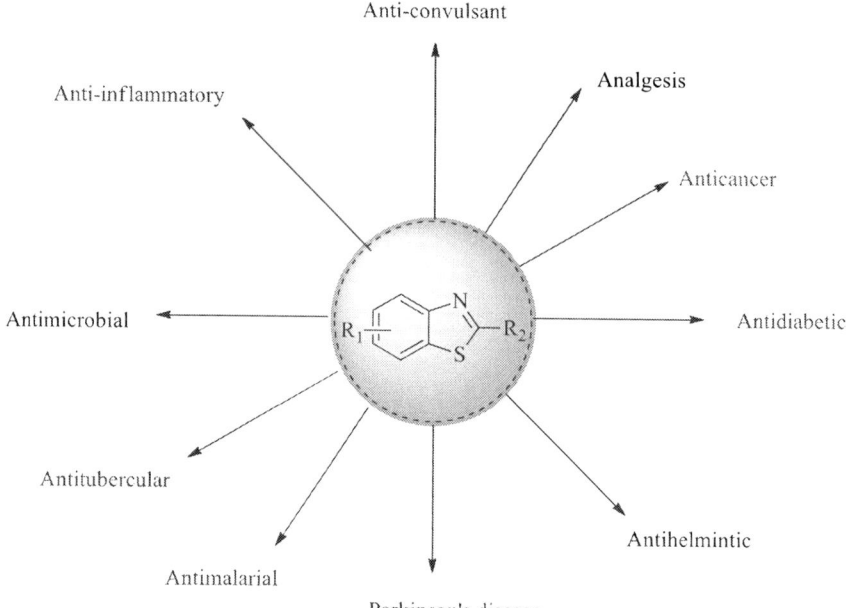

Figure 9. Biological activities of benzothiazoles.

5.1.1. Anticancer Properties of Benzothiazoles

Cancer is the second largest cause of death worldwide, extensive research has been carried out for the development of efficient chemotherapeutic agents to treat cancer. Benzothiazoles have been marked as one of the important scaffold in the cancer drug discovery and a large number of benzothiazole derivatives have demonstrated broad spectrum of anticancer properties.

2-(4-Aminophenyl)benzothiazole (**CJM 126**) and its analogs comprise a novel mechanistic class of antitumor agents [33,34]. Stevens et al. have developed phortress (**NSC 710305**) as a lead molecule from this series which is undergoing phase-I clinical trials.

2-(3,4-Dimethoxyphenyl)-5-fluorobenzothiazole (PMX 610) (Figure 10) is a derivative of 2-phenylbenzothiazole that has demonstrated effective (GI50 < 0.1 nM) and selective *in vitro* anti-proliferative activity against a panel of National Cancer Institute (NCI) 60 human cancer cell lines [35], and also exhibited notable antitumor activity against malignant cell lines

[36]. YM-201627 is another benzothiazole derivative developed as an orally active antitumor agent with selective inhibition of vascular endothelial cell proliferation [37].

Several benzothiazole based anticancer agents were found to target topoisomerases [38-42], microtubules [43-45], cytochrome P450 enzyme [46-48], rapidly accelerated fibrosarcoma (RAF) kinases, transforming growth factor-β (TGF-β), farnesyltransferase and DNA [49-54].

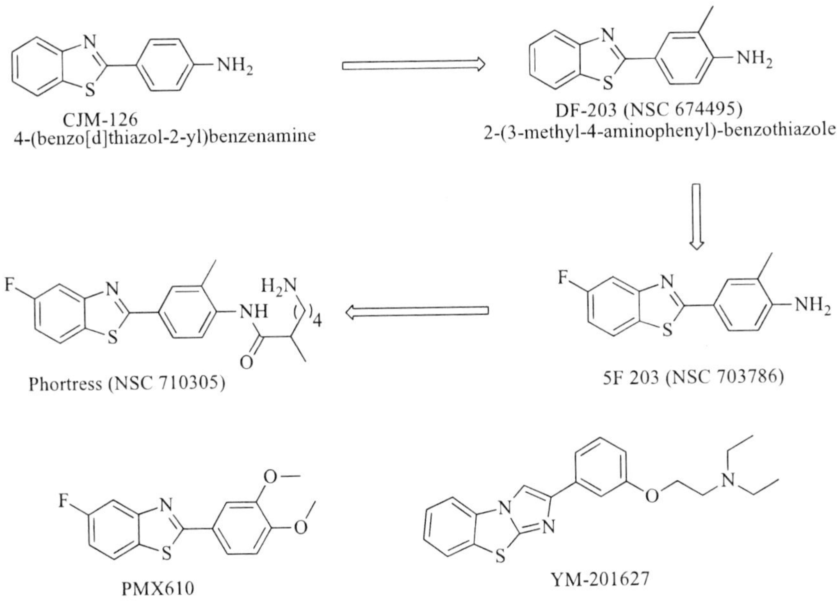

Figure 10. Anticancer benzothiazothiazole derivatives.

5.1.2. Benzothiazole as Antitubercular Agents

Benzothiazole moiety has been the privileged scaffold in tubercular drug discovery. Several benzothiazole derivatives have demonstrated potential antitubercular activity with impressive MIC values against *M. tuberculosis* H37Rv. Benzothiazole moiety has been exploited as a core site in the development of non-covalent decaprenylphosphoryl-β-D-ribose 2'-epimerase (DPrE1) inhibitors. Following benzothiazole derivatives have been identified as potential non-covalent DPrE1 inhibitors:

Wang et al. have identified benzothiazole based small molecular compound, TCA1 as non-covalent DPrE1 inhibitor with potential activity against drug-resistant and persistent tuberculosis [55]. In a parallel study, Fauzia et al. have reported the potential antitubercular activity of 1,2,3-triazole tethered benzothiazole derivatives **27** with MIC 8 μM and their molecular docking analysis revealed the non-covalent interactions with DPrE1 [56]. Binding between DPrE1 and 1,2,3-triazole conjugates of benzothiazole was confirmed from Surface Plasmon Resonance (SPR) technique [57]. Later, inspired by the structures of TCA1 and 1,2,3-triazole conjugates of benzothiazole, benzothiazolylpyrimidine-5-carboxamides **28**, were developed as non-covalent DPrE1 inhibitors by structure-based drug design approaches [58].

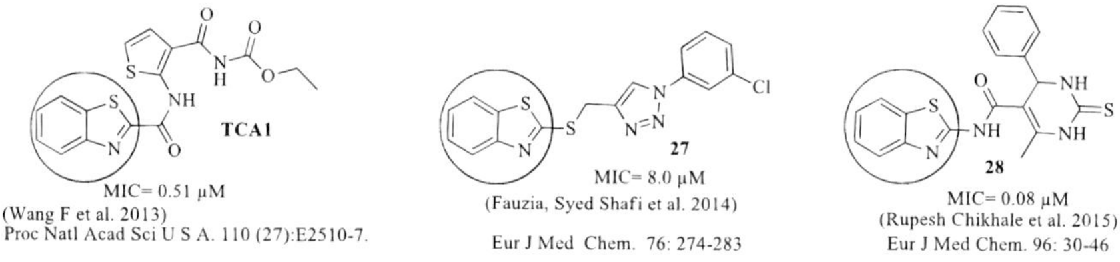

Figure 11. Non-Covalent DPrE1 inhibitors.

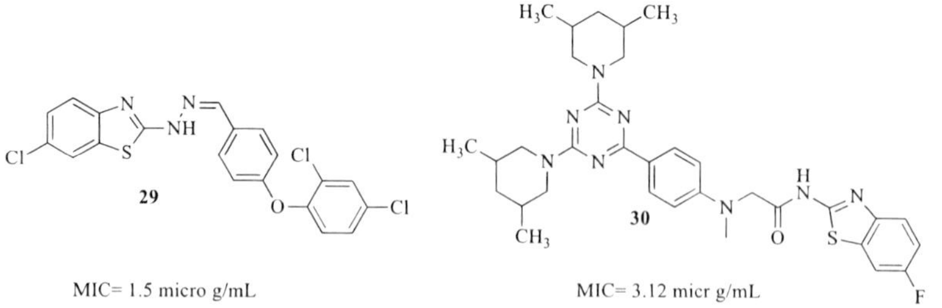

Figure 12. Benzothiazole as antitubercular agent.

A series of 2-(2-(4-aryloxybenzylidene)hydrazinyl)-benzothiazoles **29** have been synthesized using hybrid conjugation of bioactive ligands approach by combining the 2-hydrazinylbenzothiazole and 4-(aryloxy)benzaldehyde and all the synthesized compounds have shown promising anti-tubercular activity with MIC values of 1.5-29 mg/mL against

M. tuberculosis H37Rv [59]. N-Benzothiazolyl acetamide derivatives **30** displayed antitubercular activity against *Mtb H37Rv* with MIC of 3.12 μg/mL [60].

5.1.3. Benzothiazole as Anticonvulsants

Several benzothiazole derivatives have been reported to exhibit anticonvulsant and neuroprotective effects [61]. N-(Substituted benzothiazol-2-yl)amide **31** was found to exhibit significant neuroprotective effect by lowering MDA and LDH levels [62].

Benzothiazole-quinazolinone conjugate **32** displayed excellent activity against tonic seizure in the maximal electroshock (MES) model and clonic seizure in PTZ induced seizure model [63].

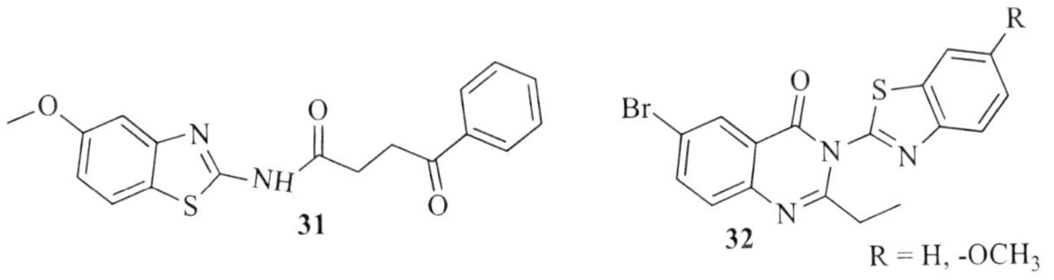

Figure 13. Benzothiazoles as anti-convulsant agents.

5.1.4. Benzothiazoles as Antidiabetic Agents

Benzothiazole derivatives have demonstrated good antidiabetic properties through various mechanisms. Compound **33** (shown in Figure 14) displayed better anti-hypoglycaemic activity than the standard drug glibenclamide in streptozotocin-induced diabetic rat model. It acts by stimulating the sulfonylurea receptor 1 (SUR1) which is a regulatory subunit of ATP-sensitive potassium channels (KATP) in pancreatic beta cells [64].

Compound **34** ((*E*)-1-(4-bromophenyl)-3-((6-ethoxybenzo[d]thiazol-2-yl)amino)-3-(methylthio)prop-2-en-1-one) exhibited good antidiabetic activity and was found to exhibit α-amylase and glucosidase inhibiting activity [65]. Compound **35,** a benzothiazole derivative of thiazolidinedione was found to be PPARγ agonist, indicating their potential as drug candidate for diabetes [66].

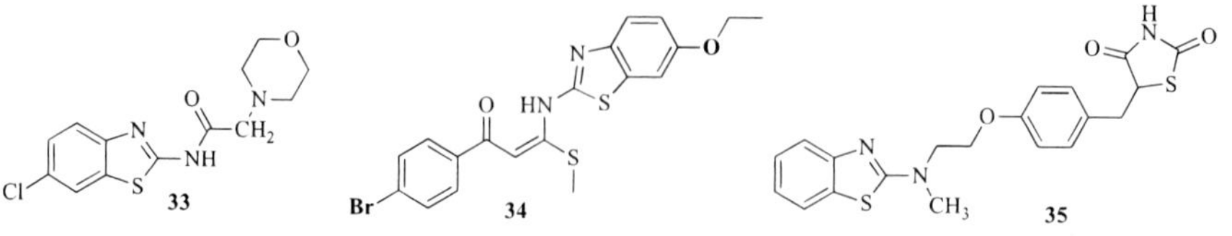

Figure 14. Benzothiazole as anti-diabetic agents.

5.1.5. Benzothiazoles as Antihelmintic Agents

2-Amino benzothiazoles having various substitutions like nitro, chloro, fluoro, bromo, methyl, ethyl, methoxy or dimethyl groups at the sixth position (compound **36**) exhibited potential antihelmintic activity [67]. Compounds **37** and **38** (in Figure 15) have demonstrated most potent antihelmintic activity compared to the standard drug albendazole in the *in vitro* assay [68].

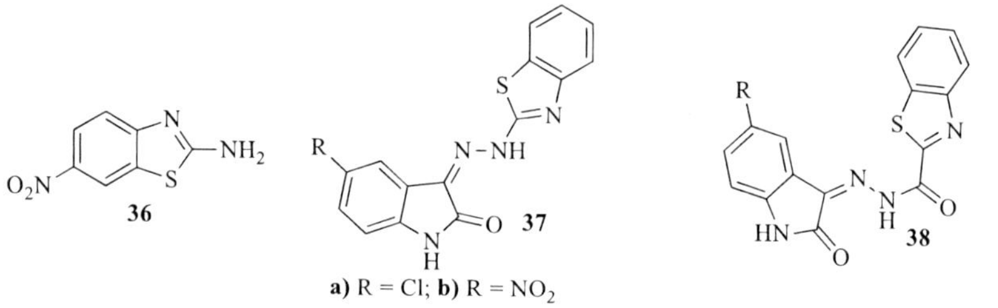

Figure 15. Benzothiazoles as anthelmintic agents.

5.1.5. Benzothiazoles as Antimalarial Agents

A series of amodiaquine analogues have been synthesized and evaluated for antiplasmodial activity against W2 and K1 chloroquine resistant strains of *Plasmodium falciparum*. Among all the synthesized compounds, compound **39** and **40** displayed potent antiplasmodial activity and interestingly inhibits W2 and K1 chloroquine resistant strains of *Plasmodium falciparum* [69]. Moreover, several derivatives of 2,6-substituted and 2,4-substituted-benzo[d]thiazoles (compound **41**) have been evaluated for the mosquito repellent activity against *Anopheles arabiensis*.

Among all, **41b-41d** (Figure 16) showed repellent activity similar to that of the positive control that is N,N-diethyl-m-toluamide (DEET) [70].

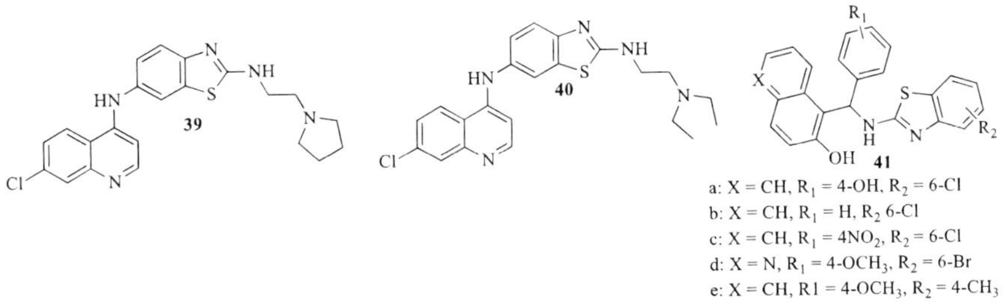

a: X = CH, R$_1$ = 4-OH, R$_2$ = 6-Cl
b: X = CH, R$_1$ = H, R$_2$ 6-Cl
c: X = CH, R$_1$ = 4NO$_2$, R$_2$ = 6-Cl
d: X = N, R$_1$ = 4-OCH$_3$, R$_2$ = 6-Br
e: X = CH, R1 = 4-OCH$_3$, R$_2$ = 4-CH$_3$

Figure 16. Benzothiazoles as anti-malarial agents.

5.1.7. Benzothiazole as Antimicrobial Agents

Various thiazolodines have demonstrated a broad spectrum of antibacterial and antifungal activity. They have shown good antibacterial activities against the both gram +ve and gram –ve bacteria. Thiazolidinone incorporated benzothiazole derivatives **42a-g** have demonstrated potential antimicrobial activity. Compound **42c** and **42d** were found to be potent as antibacterial and antifungal agents. Whereas compound **42g** exhibited potent antibacterial activity only and compound **42f** exhibited potent antifungal property as given in Figure 17 [71]. 3-(4-(Benzo[d]thiazol-2-yl)phenyl)-2-(4-methoxyphenyl)thiazolidin-4-one **43a** and 3-(4-(benzo[d]-thiazole-2-yl)phenyl-2-(4-hydroxy-phenyl)thiazolidin-4-one **43b** were found to be potent against *E. coli* and *C. albicans* with MICs ranging between 15.6-125 mg/mL [72]. Similarly, compound **44** also exhibited promising anti-microbial activity [73].

Benzothiazoles embedded with 1,2,3-triazole nucleus **45** exhibited both antibacterial and antifungal activities. However compound **45a** was found to be active against *S. aureus, E. faecalis, S. typhi, E. coli, K. pneumonia, P. aeruginosa* with MIC value 3.12 mg/mL, whereas compound **45b** displayed maximal potency against all the fungal strains such as *C. tropicalis, C. albicans, C. krusei, Cryptococcus neoformans, A. niger* and *A. fumigates* [74].

A series of 6-(benzothiazol-2-yl)pyrido[2,3-d]-pyrimidine **46** has exhibited significant antibacterial and antifungal activity against *Staphylococcus aureus, Streptococcus pyogenes, Escherichia coli, Klebsiella pneumoniae, Pseudomonas aeruginosa, Penicillium marneffei, Aspergillus flavus, Aspergillus fumigatus, Candida albicans,* etc. and were better than the standard drugs like ciprofloxacin and clotrimazole [75].

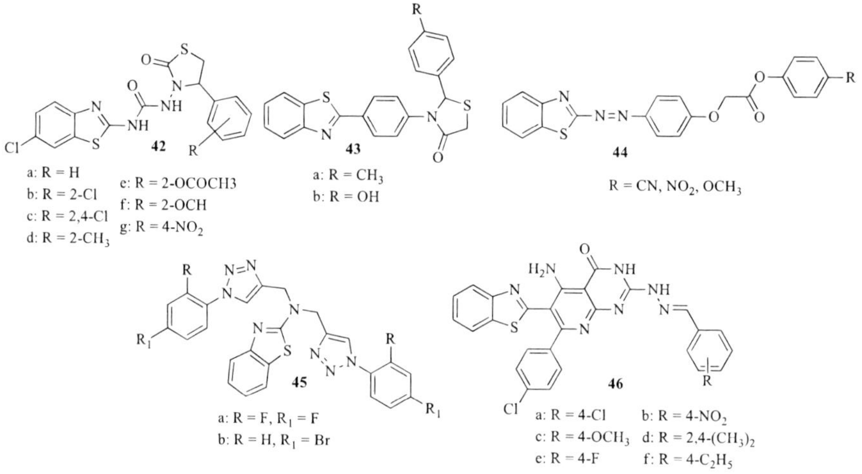

Figure 17. Benzothiazole as antimicrobial agents.

5.1.8. Benzothiazoles as Anti-Inflammatory Agents

Several benzothiazole derivatives have been explored in order to develop new class of non-steroidal anti-inflammatory drugs (NSAIDs). 2-Benzylbenzo[d]thiazole-6-sulfonamide derivatives (**47**) have shown considerable anti-inflammatory activity in carrageenan induced rat paw oedema model (Figure 18) [76]. Benzothiazoles tethered with 1,2,3-triazoles (**48**) have displayed potential anti-inflammatory activities with Cox-2 selectivity [77].

Benzothiazoles combined with thiazolidinones (**49**) exhibited anti-inflammatory as well as antifungal activities [78]. Amide derivatives of 2-aminobenzothiazole **50** exhibited potent anti-inflammatory activity with no gastrointestinal side-effects due to its non-acidic nature [79].

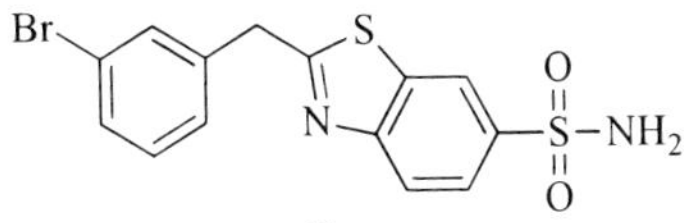

Figure 18. Benzothiazoles as anti-inflammatory agents.

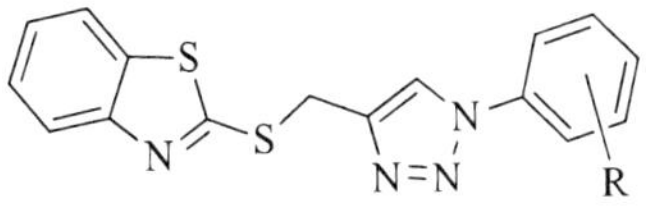

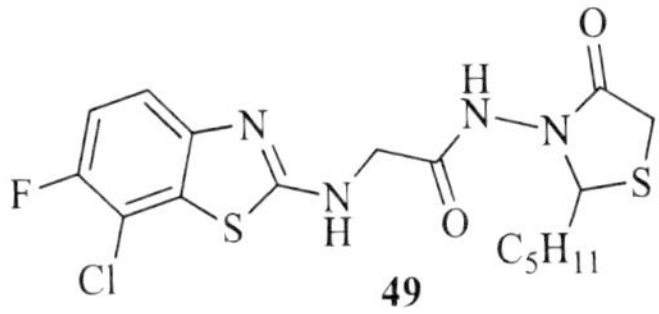

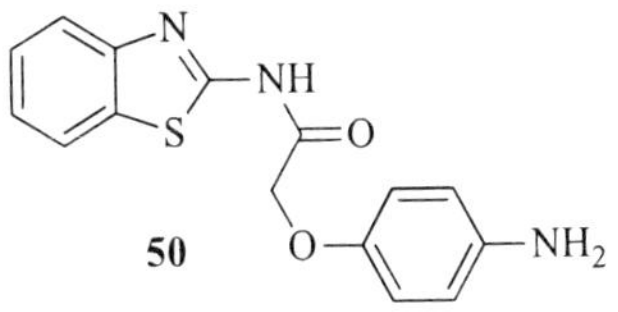

5.1.9. Benzothiazoles as Analgesic Agents

Compounds **48**, **51** and **52** (Figures 18 and 19) exhibited good analgesic activity with anti-inflammatory property [77, 80, 81].

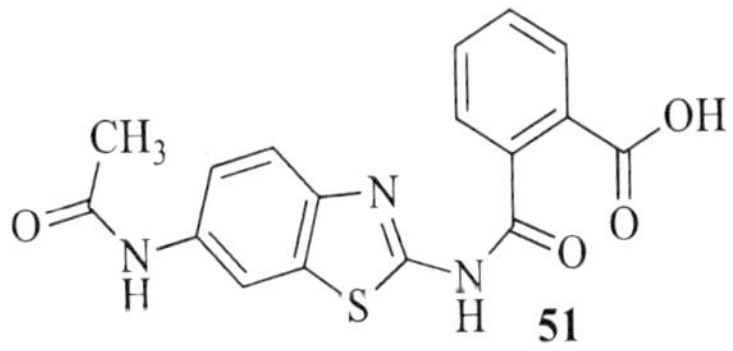

Figure 19. Benzothiazoles as analgesic agents.

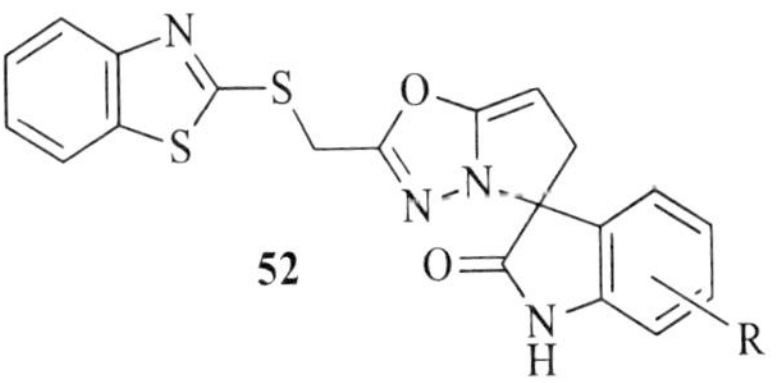

5.1.10. Miscellaneous Properties

Various 4-thiazolidinone derivatives embedded with benzothiazole moiety were synthesized and evaluated for their entomological effects (antifeedant and acaricidal activities, contact and stomach toxicities) for example compound **53** possessed good entomological effects [82]. Compound **54** exhibited reasonable anti-nematodal or schistosomicidal activity upon screening for anti-parasitic activity [83]. Compound **55** demonstrated good antipsychotic activity with muscle relaxant property as shown in Figure 20 [84].

Compound **56** exhibited potent antiulcer, anti-inflammatory, antibacterial and antitumor activities. Herein, a series of biphenylbenzothiazole-2-carboxamides have displayed diuretic potential [85].Compound **57** having substitution at 6 position of the benzothiazole ring system with electron withdrawing groups (Cl, F, Br) significantly increases urinary excretion [86]. Another benzothiazole derivative **58** was found to be α2A receptor antagonist for the treatment of conditions such as insomnia, drug addiction, pain, depression and Parkinson's disease [87]. Some of the substituted 2-aminobenzothiazole derivatives **59** showed good plant growth regulating properties equivalent to that of substances with auxin-like growth promoting activity [88].

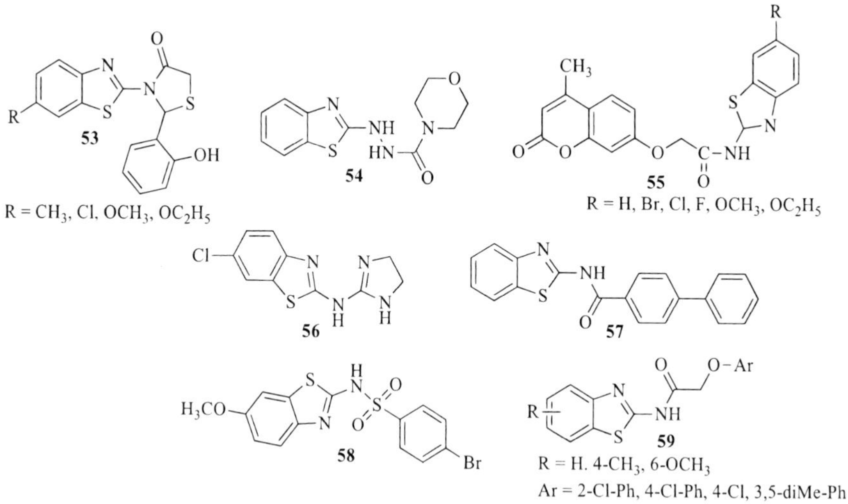

Figure 20. Benzothiazole analogs for miscellaneous properties.

5.2. Industrial Applications of Benzothiazole Derivatives

Apart from its biological activity, benzothiazoles act as multiple functionality heterocycle for the starting material of other industrial chemicals. Basically, benzothiazoles are used as biocides and corrosion inhibitors in the manufacturing industry.

Among all the benzothiazoles, 2-mercaptobenzothiazole **11** (2-SH-BTH) and 2-morpholinothiobenzothiazole **60** (Figure 21) are utilized as vulcanization accelerators in rubber manufacturing industry mainly in automobile tires, which is added about 1% of weight to the rubber.

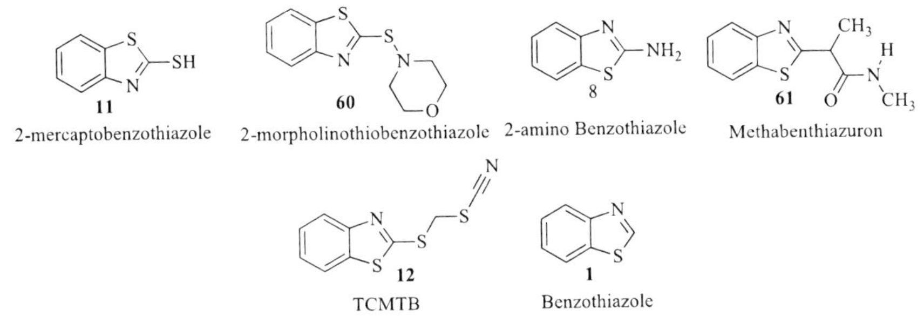

Figure 21. Benzothiazole derivative as industrial chemicals.

They are also used as corrosion inhibitors in cooling liquids and antifreezes. Moreover, they are consumed as constituents of azo dyes (2-amino benzothiazole) apart from photosensitizers and UV light stabilizers in textiles and plastics. Various 2-substituted benzothiazoles are utilized as algicides, herbicides (methabenthiazuron), slimicides in paper & pulp industry and fungicides (compounds **12** and **1**) in lumber & leather industry [89a].

TCMTB is used as wideband microbicide, paint fungicide and paint gallicide [89b]. It is used in leather preservation [89c], protection of paper products, wood preservatives and against germs in industrial water. TCMTB is being used as a fungicide for seed dressing in cereals, safflower, cotton and sugar beet. It is also used to deal with fungal problems while extracting hydrocarbons *via* fracking [89d].

5.3. Applications of Benzothiazoles in Organic Synthesis

Synthetically benzothiazoles are also useful in chemical transformations. Primary amines have been transformed to aldehydes or ketones through imines by reacting with benzothiazole-2-carboxaldehyde as shown below (Figure 22).

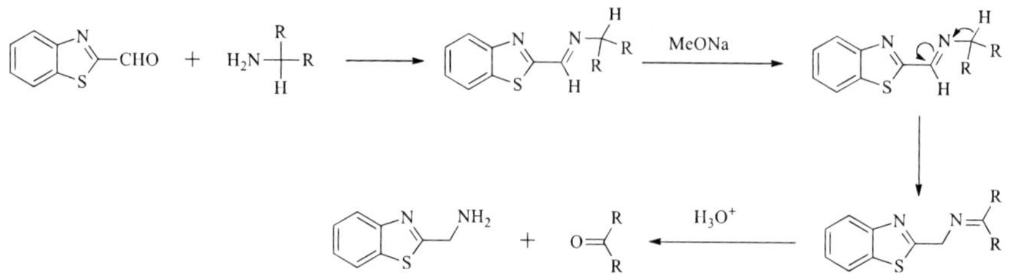

Figure 22. Transformation of primary amines to carbonyl compounds using benzothiazoles.

An attractive method for the preparation of ketones has been developed by using 2-mercaptobenzothiazoles as shown below (Figure 23).

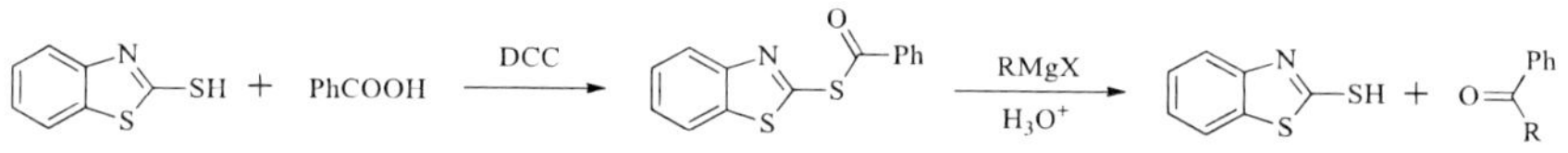

Figure 23. Preparation of ketones by using 2-mercaptobenzothiazoles.

6. TOXIC EFFECT OF VARIOUS BENZOTHIAZOLE DERIVATIVES

Even though benzothiazoles have a wide range of applications, they exert a variety of in *vivo* or *in vitro* toxic effects [89a]. Some of them are as follows:

6.1. Modulation of Thyroid Hormone

Some of the benzothiazole derivatives such as 2-mercaptobenzothiazole (2-SH-BTH),5-chloro-2-mercaptobenzothiazole, 2-aminobenzothiazole (2-NH_2-BTH),2-hydroxybenzothiazole (2-OH-BTH) and 2-(methylthio) benzothiazole (2-Me-S-BTH) cause the modulation of thyroid hormone. It has reported that these derivatives inhibit the thyroid peroxidase in porcine thyroid glands in *in vitro* with inhibitory order of 2-SH-BTH = 5-chloro-2-

mercapto-BTH > 2-NH$_2$-BTH > BTH, with IC50 values of 12, 13, 1200, and 10000 µM, respectively [90].

6.2. Genotoxicity and Cytotoxicity

Benzothiazole (BTH), 2-chloro-BTH, 2-bromo-BTH, 2-fluoro-BTH, 2-Me-BTH, 2-SH-BTH, 2-NH$_2$-BTH, 2-OH-BTH and 2-Me-S-BTHBTH, 2-OH-BTH and 2-NH$_2$-BTH showed cytotoxicity to *S. typhimurium* 420 TA1535/pSK1002 strain. With the exception of BTH and 2-fluoro-BTH, all the tested BTs were more cytotoxic to bacteria than to human carcinoma cells. Further on, the genotoxicity of the nine BTs were screened by SOS/umu test on *S. typhimurium* TA1535/pSK1002. All the tested the BTs except for BTH, 2-fluoro- BTH and 2-SH-BTH induced genotoxicity by damaging DNA and chromosomes [91].

6.3. Neurotoxicity

It is found that coffee extracts induces neurotoxicity by changing γ-aminobutyric acid type A (GABA$_A$) receptors at high doses (0.5-0.8 µL/mL) expressed in *Xenopus* oocytes, which inhibited the GABA in dose-dependent manner. BTH is one of the components in coffee extracts which significantly potentiates the GABA responses [92].

6.4. Allergic and Dermatitis Reactions

Various allergic and dermatitis reactions have been reported in humans after topically exposure to benzothiazole as early as 1931. Similarly, 2-SH-BTH induced dermatitis and skin sensitizationin humans and rodents [93]. Irritation of nose and throat has also found in asphalt-rubber workers due to their involvement in laying pavements which was thought to be due to the presence of BTH in asphalt-rubber [94]. Various other allergic reactions with sensory and pulmonary irritation were illustrated in *ex vivo* lymph node cell models and mice due to BTH.

6.5. Hepatotoxicity

Oral exposure of mice to BTH leads to increase in the activity of phase-I metabolic and conjugation enzymes [95]. Similarly, oral exposure of BTH followed by intraperitoneal injection with acetaminophen enhanced the serum alanine aminotransferase activities. It was suggested that BTH ultimately causes the hepatotoxicity by inducing phases II metabolizing enzymes.

6.6. Congenital Malformations

Studies have shown that 2-SH-BTH induces fetotoxic and teratogenic effects in chicken embryos upon injecting 0.1-2 µmol/egg into the air sac for 11 days [96]. Congenital malformations including defects in eye, neck and back, as well as open coelom, were seen in 20% of the chicken embryos [97].

6.7. Occupational Exposure

Various epidemiological studies have reported that the rubber industry workers have a high risks for a variety of cancers including bladder cancer, lung cancer and leukemia. 2-SH-BTH derivative is one of the benzothiazoles used in rubber industry. Higher rates of bladder cancer have occurred in the chemical factory of northern Wales due to exposure of vulcanization inhibitors, accelerators, antioxidants, and other specialty chemicals. Various aromatic amines, in particular, phenyl-β-naphthylamine and *o*-toluidine along with 2-SH-BTH also leads to bladder cancer [98]. About 360 male production workers exposed to 2-SH-BTH had mortality and cancer morbidity (bladder and intestine) at the chemical factory in northern Wales. Similarly, a cohort of 600 workers at a nitro plant exposed to 2-SH-BTH and *p*-aminobiphenyl in West Virginia, USA also showed higher rate of bladder cancer.

6.8. Ecotoxicity in Aquatic Organisms

BTs cause the ecotoxicity in aquatic organisms as a result of industrial discharges roadside runoff and wastewater. Due to its high water solubility, sorption and bioaccumulation potentials of benzothiazoles are thought to be low in water column. BTH decreased the activities of digestive enzymes such as protease, lipase, α-amylase and trehalase and reduced nutrient accumulation (protein, lipid, carbohydrate, and trehalose) in the larvae of *Bradysia odoriphaga*. 2-SCNMeS-BTH was generally the most toxic derivative among benzothiazoles which induced ecotoxicity in fish, invertebrates and aquatic plants [88].

CONCLUSION

Benzothiazole derivatives have received much attention in medicinal chemistry due to their wide range of biological, pharmacological and industrial applications. Riluzole and pramipexole are some of the drugs bearing benzothiazole scaffold that are used in the treatment of amyotrophic lateral sclerosis and Parkinson's disease respectively. Benzothiazole derivatives have a large number of applications in material chemistry as they were used as vulcanization accelerators, dyes, corrosion inhibitors, insecticides, fungicides, herbicides, algicides, food flavoring agents, nonlinear optics (NLO) and ultraviolet (UV) light stabilizers over the decades. Benzothiazole alkaloids are relatively rare in nature and are present in a range of marine or terrestrial natural compounds. Luciferin is the well known example of naturally occurring benzothiazole that is found in the organisms that exhibit biolumininescence. Even though benzothiazoles displays diverse applications, some of the benzothiazole derivatives are highly toxic. Apart from their biological, pharmaceutical and industrial applications, their toxic effects have also been discussed in this chapter.

REFERENCES

[1] Green, A. G. and Perkin, A. G. (1903). *Journal of the Chemical Society*, 70, 1207. Accessed October 12, 2019.

[2] Justus, K. L. (1967). Diaminobenzobisthiazoles and Related Compounds. *Journal of the Chemical Society*, 2212-20. Accessed October 12, 2019. doi: 10.1039/J39670002212

[3] Kiprianov, A. I. and Mikhailenko, F. A.(1967). Synthesis and structures of isomeric benzobisthiazoles. *Chemistry of Heterocyclic Compounds*, 3, 205-09. Accessed October 12, 2019. https://link.springer.com/article/10.1007/BF01172550.

[4] Samuel, L. S., Robert, J. C., Nicholas, J. C. and Raymond, E. (1968). Synthesis and proof of structure of 2,6-diamino-benzo[1,2-d:4,5-d']bisthiazole. *The Journal of Organic Chemistry*, 33, 2132-33. Accessed October 12, 2019. doi: 10.1021/jo01269a102.

[5] Mukherjee, R. (2018). A Review on the Synthesis and Use of Benzobisthiazoles: An Important Class of Heterocycles. *International Journal of Scientific Research and Reviews*, 7, 1050-69. Accessed October 12, 2019. url: file:///C:/Users/DR/Downloads/pdf_1599.pdf.

[6] Anjou, K. and Von, E. S. (1967). The aroma of cranberries. II. Vaccinium macrocarpon Ait. *Acta Chemica Scandinavica*, 21, 2076-82. Accessed October 12, 2019. doi: 10.3891/acta.chem.scand.21-2076

[7] Baines, D. A., Faulkes, C. G., Tomlinson, A. J. and Ning, P. C. Y. K. (1989). *US Patent 4 818 535*, 1989.Accessed October 12, 2019.

[8] Bruno, F., Franco, M., Denis, B., Giorgio, N. and Giuseppe, V. (2007). Aging Effects and Grape Variety Dependence on the Content of Sulfur Volatiles in Wine. *Journal of Agricultural and Food Chemistry*, 55, 10880-87. Accessed October 12, 2019. doi:10.1021/JF072145W.

[9] Vincenzo, B., Marco, N., Roberto, F. and Domenico, R. (2000). Analysis of Benzothiazole in Italian Wines Using Headspace Solid-Phase Microextraction and Gas Chromatography–Mass Spectrometry. *Journal of Agricultural and Food Chemistry*, 48, 1239-42. Accessed October 12, 2019. doi:10.1021/JF990634T.

[10] Pérez-Coello, M. S., Sanz, J. and Cabezudo, M. D. (1998). Gas chromatographic-mass spectrometric analysis of volatile compounds in oak wood used for ageing of wines and spirits. *Chromatographia*, 47, 427-32. Accessed October 12, 2019. doi:10.1007/BF02466474.

[11] Otto, G. V., Peter, W. and Peter, H. (1975). New volatile constituents of black tea aroma. *Journal of Agricultural and Food Chemistry*, 23, 999-1003. Accessed October 12, 2019. doi: 10.1021/JF60201A03.

[12] Richard, M. S. and King, A. J. D. (1982). Identification of some volatile constituents of *Aspergillus clavatus*. *Journal of Agricultural and Food Chemistry*, 30, 786-90. Accessed October 12, 2019. doi:10.1021/JF00112A044.

[13] Kozasa, T., Suzuki, K., Tsunoda, N., Tanaka, K., Yoneda, T. and Hirasawa, M. (1998). *Japanese Patent 63030493-A*, 1988.

[14] Yaginuma, S., Asahi, T., Takada, M., Hayashi, M. and Mizuno, K. (1989). *Japanese Patent 1181793-A*, 1989.

[15] Barbara, B. and McElroy, W. D. (1957). "The preparation and properties of crystalline firefly luciferin." *Archives of Biochemistry and Biophysics*, 72, 358-68. Accessed October 12, 2019. doi:10.1016/0003-9861(57) 90212-6.

[16] Emil, H. W., Frank, M. and George, F. F. (1963). "The Structure and Synthesis of Firefly Luciferin." *Journal of the American Chemical Society*, 85, 337-43. Accessed October 12, 2019. doi:10.1021/JA00886A019.

[17] Andrea, A. S., John, H. C. and Fred, L. S. (1991). "Benzothiazoles from a putatitve bacterial symbiont of the marine sponge *Tedania ignis*. *Tetrahedron Letters*, 32, 4847-48. Accessed October 12, 2019. doi:10.1016/S0040-4039(00)93476-2.

[18] Hosokawa, N., Naganawa, H., Hamada, M., Iinuma, H., Takeuchi, T., Tsuchiya, K. S. and Hori, M. (2000). New triene-ansamycins,

thiazinotrienomycins F and G and a diene-ansamycin, benzoxazomycin. *The Journal of antibiotics*, 53, 886-94. Accessed October 13, 2019. doi:10.7164/antibiotics.53.886.

[19] Takeuchi, T., Hori, M., Hamada, M., Osanawa, H., Iinuma, H. and Hosokawa, N. (2001). *Japanese Patent 2001199988-A*,2001.

[20] (a) Toshiaki, M. and Takahiro, I. (2008). General and practical synthesis of Benzothiazoles. *Pure and applied chemistry*, 80, 707–15. Accessed October 13, 2019.doi: 10.1351/pac200880040707.

(b) Neelam, P., Prajapati, R. H. V., Mayuri, A. B. and Hitesh, D. Patel. (2014). Patel. Recent advances in the synthesis of 2-substituted Benzothiazoles: A Review. *Royal Society of Chemistry Advance*, 4, 60176-208. Accessed October 13, 2019. doi:10.1039/C4RA07437H

(c) Victor, F., Raisa, d. R. Reis., Claudia, R. B. G. and Thatyana, R. A. V. (2012). Chemistry and Biological Activities of 1,3-Benzothiazoles. *Mini-Reviews in Organic Chemistry*, 9, 44-53. Accessed October 13, 2019. doi:10.2174/157019312799079929.

(d) Gill, R. K.., Rawal, R. K., and, Bariwal, J. (2015). Recent Advances in the Chemistry and Biology of Benzothiazoles. *Archiv der Pharmazie*, 348, 155–78. Accessed October 13, 2019.
doi: 10.1002/ardp.201400340.

[21] Pratap, U. R., Mali, J. R., Jawale, D. V. and Mane, R. A. (2009). Bakers' yeast catalyzed synthesis of benzothiazoles in an organic medium. *Tetrahedron Letters*, 50, 1352–54. Accessed October 13, 2019. doi:10.1016/j.tetlet.2009.01.032.

[22] Hamid, A., Fazaeli, R., Fazaeli, Nahid., Ahmad, R. M., Hamid, J. N., Alizadeh, M. and Ginous, E. (2009). Facile route for the synthesis of benzothiazoles and benzimidazoles in the presence of tungstophosphoric acid impregnated zirconium phosphate under solvent-free conditions. *Heteroatom Chemistry*, 20, 202-07. Accessed October 13, 2019. doi: 10.1002/hc.20534.

[23] Hong, Y. G., Ji, C. L. and You, L. S. (2009). A simple and efficient synthesis of 2-substituted benzothiazolescatalyzed by H_2O_2/HCl. *Chinese Chemical Letters*, 20, 1408–10. Accessed October 13, 2019. doi: 10.1016/j.cclet.2009.06.037.

[24] Kumar, A., Sharma, S. and Maurya R. A. (2010). Bienzymatic synthesis of benzothia/(oxa)zoles in aqueous medium. *Tetrahedron Letters*, 51, 6224-26. Accessed October 13, 2019. doi: 10.1016/j.tetlet.2010.06.12.

[25] Xiao, L. Y., Chun, M. X., Shao, M. L., Jiu, X. C., Jin, C. D., Hua, Y. W. and Wei, K. S. (2010). Eco-friendly synthesis of 2-substituted benzothiazolescatalyzed by cetyltrimethyl ammonium bromide (CTAB) in water. *Journal of the Brazilian Chemical Society*, 21, 37-42. Accessed October 13, 2019.
url: http://www.scielo.br/pdf/jbchs/v21n1/07.pdf.

[26] Hideaki, H., Kimihito, M., Miwa, K. and Mitsuaki, K. (2006). Combinatorial synthesis of benzothiazoles and benzimidazoles using a traceless aniline linker. *Journal of Combinatorial Chemistry*, 8, 462-63. Accessed October 14, 2019. doi: 10.1021/cc0600472.

[27] (a)Jacobson, P. (1886). Ueberbildung von anhydroverbindungen desorthoamidophenylmercaptans aus thioaniliden. *European Journal of Inorganic Chemistry*, 19, 1067-70. Accessed October 14, 2019. doi:https: 10.1002/cber.188601901239.
(b)Malcolm, F. G. S., Carol, J. M., Peter, L., Peter, A., Audrey, R. and Donna, E. D. (1994). Structural studies on bioactive compounds. 23. Synthesis of polyhydroxylated 2-phenylbenzothiazoles and a comparison of their cytotoxicities and pharmacological properties with genistein and quercertin. *Journal of Medicinal Chemistry*, 37, 1689-95. Accessed October 14, 2019. doi: 10.1021/jm00037a020.

[28] Nadale, K. D. and Yvette, A. J. (2007). Iodine-mediated cyclisation of thiobenzamides to produce benzothiazoles and benzoxazoles. *Tetrahedron*, 63, 10276-81. Accessed October 14, 2019. doi: 10.1016/j.tet.2007.07.076.

[29] Takahiro, I. and Toshiaki, M. (2007). A novel practical synthesis of benzothiazoles via Pd catalyzed thiol cross-coupling. *Organic Letters*, 9, 3687-89. Accessed October 14, 2019. doi: 10.1021/ol7015737.

[30] Chun-Lin, L., Xing-Guo, Z., Ri-Yuan, T., Ping, Z. and Jin-Heng, Li. (2010). Copper-catalyzed thiolation annulations of 1,4-dihalides with sulfides leading to 2- trifluoromethylbenzothiophenes and

benzothiazoles. *The Journal of Organic Chemistry*, 75, 7037-40. Accessed October 14, 2019. doi:10.1021/jo101675f.

[31] Zhu, J., Chen, Z., Xie, H., Li, S. and Wu, Y. (2010). A general and straightforward method for the synthesis of 2-trifluoromethylbenzothiazoles. *Organic Letters*, 12(10), 2434-36. Accessed October 16, 2019. doi: 10.1021/ol1006899.

[32] (a) Douglas, A. H., Gregory, T. B. and Mark, L. S. (2003). The combinatorial synthesis of bicyclic privileged structures or privileged substructures. *Chemical Reviews*, 103, 893-930. Accessed October 16, 2019. doi: 10.1021/cr020033s

(b) Robert, W. D., Kevin, S. C., Mitchell, S. A., James, W. D. and Douglas, A. P. (2004). Privileged structures: applications in drug discovery. *Combinatorial Chemistry and Highthroughput Screening*, 7, 473-94. Accessed October 16, 2019. doi: 10.2174/138620 7043328544.

(c) Roland, E. D. (1996). Discovery of enzyme inhibitors through combinatorial chemistry. *Molecular Diversity*, 2, 223-36. Accessed October 16, 2019. https://link.springer.com/content/pdf/10.1007% 2FBF01715638.pdf.(d) Rangappa, S. K., Mahadeo, R. P., Siddappa, A. P. and Srinivasa, B. (2014). A comprehensive review in current developments of benzothiazolebased molecules in medicinal chemistry. *European Journal of Medicinal Chemistry*, 89, 207-51. Accessed October 16, 2019.
doi: 10.1016/j.ejmech.2014.10.059.

[33] Chee, O. L., Gaskell, M., Elizabeth, A. M., Robert, T. H., Peter, B. F., Bibby, M. C., Patricia, A. Cooper., John, A. D., Tracey, D. B. and Malcolm, F. G. Steven. (2003). Antitumour 2-(4- aminophenyl) benzothiazoles generate DNA adducts in sensitive tumour cells *in vitro* and *in vivo British Journal of Cancer*, 88, 470-77. Accessed October 18, 2019. doi: 10.1038/sj.bjc.6600719.

[34] Tracey, D. B., Wrigley, S., Dong, F. S., Robert, J. S., Kenneth, D. P. and Malcolm, F. G. S. (1998). 2- (4-Aminophenyl)benzothiazoles: novel agents with selective profiles of *in vitro* antitumour activity. *British Journal of Cancer*, 77, 745-52. Accessed October 18, 2019.

doi: 10.1038/bjc.1998.122.

[35] Boon, S. T., Kai, H. T., Ashwin, M., Nirmal, R., Heng, L. C., Tracey, D. B., Malcolm, F. S. and Chee, O. L. (2011). CYP2S1 and CYP2W1 mediate 2-(3,4- dimethoxyphenyl)-5-fluorobenzothiazole (GW-610, NSC 721648) sensitivity in breast and colorectal cancer cells. *Molecular Cancer Therapeutics*, 10, 1982-92. Accessed October 18, 2019. doi: 10.1158/1535-7163.MCT-11-0391.

[36] Yaseen, A. A., Haitham, H. A., Saeed, B., Ihsan, H. J., Mohammad, O. B., Najim, A. A. M., Tahsin, A. K., Paolo, L. C., Busonera, B., S, T. and Loddo, R.(2008). Synthesis and *in vitro* antiproliferative activity of new benzothiazole derivatives. *ARKIVOC*, 15, 225-38. Accessed October 18, 2019.http://www.arkat-usa.org/get-file/26406/.

[37] Amino, N., Ideyama, Y., Yamano, M., Kuromitsu, S., Tajinda, K., Samizu, K., Matsuhisa, A., Kudoh, M. and Shibasaki, M. (2006). YM-201627: an orally active antitumor agent with selective inhibition of vascular endothelial cell proliferation. *Cancer Letters*, 238, 119-27. Accessed October 20, 2019. doi: 10.1016/j.canlet.2005.06.037.

[38] Kamal. A, Ashwini Kumar, B., Suresh, P., Shankaraiah, N. and Shiva Kumar, M.(2011). An efficient one-pot synthesis of benzothiazolo-4b-anilino-podophyllotoxincongeners: DNA topoisomerase-II inhibition and anticancer activity. *Bioorganic Medicinal Chemistry Letters*, 21, 350-53. Accessed October 20, 2019. doi: 10.1016/j.bmcl.2010.11.002.

[39] Suk, J. C., Hyen, J. P., Sang, K. L., Sang, W. K., Han, G. and Hea, Y. P. C. (2006). Solid phase combinatorial synthesis of benzothiazoles and evaluation of topoisomerase IIinhibitory activity. *Bioorganic and Medicinal Chemistry*, 14, 1229-35. Accessed October 20, 2019. doi: 10.1016/j.bmc.2005.09.051.

[40] Cigdem, K. O., Betul, T. G., Foto, E., Yildiz, I., Diril, N., Aki, E. and Yalcin, I. (2013). Benzothiazole derivatives as human DNA topoisomerase IIα Inhibitors. *Medicinal Chemistry Research*, 22, 5798-808. Accessed October 20, 2019. doi: 10.1007/s00044-013-0577-5.

[41] Pinar, A., Yurdakul, P., Yildiz, I., Ozlem, Temiz-Arpaci., Leyla, N., Acan, E. A. and Yalcin, I. (2004). Some fused heterocyclic compounds

as eukaryotic topoisomerase II inhibitors. *Biochemical and Biophysical Research Communications*, 317, 670-74. Accessed October 22, 2019. doi: 10.1016/j.bbrc.2004.03.093.

[42] Abdel-Aziz, M., Matsuda, K., Otsuka, M., Uyeda, M., Okawara, T. and Suzuki, K. (2004). Inhibitory activities against topoisomerase I & II by polyhydroxybenzoylamide derivatives and their structure-activity relationship. *Bioorganic Medicinal Chemistry Letters*, 14, 1669-72. Accessed October 22, 2019. doi: 10.1016/j.bmcl.2004.01.060.

[43] Kamal A., Mallareddy, A., Suresh, P., Thokhir, B. Shaik., Lakshma Nayak, V., Kishor, C.,Rajesh, V. C. R. N. C. Shetti., Sankara Rao, N., Jaki, R. Tamboli., Ramakrishna, S. and Anthony, Addlagatta. (2012). Synthesis of chalcone-amidobenzothiazole conjugates as antimitotic and apoptotic inducing agents. *Bioorganic Medicinal Chemistry*, 20, 3480-92. Accessed October 22, 2019. doi: 10.1016/j.bmc.2012.04.010.

[44] Kamal A., Ratna Reddy, C. and Prabhakar, S.(2012). *2-Phenyl benzothiazole linked imidazole compounds as potential anticancer agents*. WO 2012110959 A1.

[45] Kamal A., Sultana, F., Janaki Ramaiah, M., Srikanth, Y. V. V., Viswanath, A., Chandan, Kishor., Pranjal, Sharma., Pushpavalli, S. N. C. V. L., Addlagatta, A. and Manika, P. Bhadra. (2012). 3- Substituted 2-phenylimidazo[2,1-b]benzothiazoles: synthesis, anticancer activity, and inhibition of tubulin polymerization. *Chem Med Chem*, 7, 292-300. Accessed October 22, 2019. doi: 10.1002/cmdc.201100511.

[46] Alessandro, S., Martin, F. and Rolf, W. H. (2012). Optimization of hydroxybenzothiazoles as novel potent and selective inhibitors of 17b-HSD1. *Journal of Medicinal Chemistry*, 55, 2469-73. Accessed October 24, 2019. doi: 10.1021/jm201711b.

[47] Hutchinson, I., Mei-Sze, C., Helen, L. B., Trapani, V., Tracey, D. B., Andrew, D. W. and Malcolm, F. G. S. (2001). Antitumor benzothiazoles. Synthesis and *in vitro* biological properties of

fluorinated 2-(4-aminophenyl)benzothiazoles. *Journal of Medicinal Chemistry*, 44, 1446-55. Accessed October 24, 2019. doi: 10.1021/jm001104n.

[48] Kashiyama, E., Hutchinson, I., Mei-Sze, C., Sherman, F. S., Lawrence, R. P., Kaur, G., Edward, A. S., Tracey, D. B., Andrew, D. W. and Malcolm, F. G. S. (1999). Antitumor benzothiazoles. 8.[1] Synthesis, metabolic formation, and biological properties of the C- and N-oxidation products of antitumor 2-(4-aminophenyl)-benzothiazoles. *Journal of Medicinal Chemistry*, 42, 4172-84. Accessed October 24, 2019. doi: 10.1021/jm990104o.

[49] Kamal A., Yellamelli, V. V. Srikanth., Naseer, A. K. Mohammed., Farheen, S. Mohammad and Kashi, R. Methuku. (2012). Recent advances on structural modifications of benzothiazoles and their conjugate systems as potential chemotherapeutics. *Expert Opinion on Investigational Drugs*, 21, 619-35. Accessed October 24, 2019. doi: 10.1517/13543784.2012.676043.

[50] Dubey, R., Prabhat, K. S., Pawan, K. B., Bhattacharya, S. and Narayana, S. M. (2006). 2- (4-Aminophenyl) benzothiazole: a potent and selective pharmacophore with novel mechanistic action towards various tumour cell lines. *Mini Reviews in Medicinal Chemistry*, 6, 633-37. Accessed October 26, 2019. doi: 10.2174/138955706777435706.

[51] Singh, M. and Singh, S. K. (2014). Benzothiazoles: how relevant in cancer drug design strategy? *Anti-Cancer Agents in Medicinal Chemistry*, 14, 127-46. Accessed October 28, 2019. doi: 10.2174/18715206113139990312.

[52] Amir, M. and Hassan, M. Z. (2013). Functional roles of benzothiazole motif in antiepileptic drug research. *Mini Reviews in Medicinal Chemistry*, 13, 2060-75. Accessed October 28, 2019. doi: 10.2174/1389557513666131119203036.

[53] Tracey, D. B. and Westwell, A. D. (2004). The development of the antitumour benzothiazole prodrug, Phortress as a clinical candidate. *Current Medicinal Chemistry*, 11, 1009-21. Accessed October 28, 2019. doi:10.2174/0929867043455530.

[54] Livio, R., Sandra, K. P., Raja, N., Sabine, D., Charles, P. C., Ivana, R., Marie, H. D. C., Kresimir, P., Vesna, T. K. and Grace, K. Z. (2013). New anticancer active and selective phenylene-bisbenzothiazoles: synthesis, antiproliferative evaluation and DNA binding. *European Journal of Medicinal Chemistry*, 63, 882-91. Accessed October 28, 2019.doi:10.1016/j.ejmech.2013.02.026.

[55] Feng, W., Dhinakaran, S., Rajkumar, H., Jianing, W., Sarah, M. B., Brian, W., Ahmad, I., Yang, P., Zhang, Y., John, K., Morad, H., Stanislav, H., Claudia, T., Zhenkun, M., Takushi, K., Khisi, E. M., Scott, F., Arnab, K. C., Kai, J., Katarina, M., Gurdyal, S. B., Klaus, F., Scott, H. R., Whitney B. S., John, R. W., William, R. J. and Peter, G. Schultz. (2013). Identification of a small molecule with activity against drug-resistant and persistent tuberculosis. *Proceedings of the National Academy of Sciences of the United States of America*, 110, E2510-17. Accessed October 28, 2019. doi: 10.1073/pnas. 1309171110.

[56] Mir, F., Shafi, S., Zaman, M. S., Nitin, P. Kalia., Rajput, V. S., Mulakayala, C., Mulakayala, N., Khan, I. A. and Mohammad, S. A. (2014). Sulfur rich 2-mercaptobenzothiazole and 1, 2, 3-triazole conjugates as novel antitubercular agents. *European Journal of Medicinal Chemistry*, 76, 274-283. Accessed October 28, 2019. doi: 10.1016/j.ejmech.2014.02.017.

[57] Sumita, K., Vipin, K. K., Syed, S. and Ajay, K. S. (2017). Structural and inhibition analysis of novel sulfur-rich 2-mercaptobenzothiazole and 1,2,3-triazole ligands against Mycobacterium tuberculosis DprE1 enzyme. *Journal of Molecular Modelling*, 23, 241. Accessed October 28, 2019. doi: 10.1007/s00894-017-3403-z.

[58] Rupesh, C., Sunil, M., Ramavath, B., Ratnadeep, B., Bhargavi, G., Nazira, K., Rajasekharan, M. V., Anant, P. and Pramod, K. (2015). Development of selective DprE1 inhibitors: design, synthesis, crystal structure and antitubercular activity of benzothiazolylpyrimidine-5-carboxamides. *European Journal of Medicinal Chemistry*, 96, 30–46. Accessed October 28, 2019. doi:10.1016/j.ejmech.2015.04.011.

[59] Vikas, N. T., Kumar, V., Bairwa, K. S. and Anirudh, B. (2012). Novel 2-(2-(4- aryloxybenzylidene) hydrazinyl)benzothiazole derivatives as antituberculars agents. *Bioorganic & Medicinal Chemistry Letters*, 22, 649-652. Accessed October 28, 2019. doi: 10.1016/j.bmcl.2011.10.064.

[60] Patel, A. B., Patel, R. V., Kumari, P., Dhanji, P. R. and Kishor, H. C. (2013). Synthesis of potential antitubercular and antimicrobial s-triazine-based scaffolds via Suzuki cross-coupling reaction. *Medicinal Chemistry Research*, 22, 367-81. Accessed November 02, 2019. doi: 10.1007/s00044-013-0839-2.

[61] (a) Ajeet. and Kumar, A. (2013). Designing of Hybrid form of Benzothiazole-quinazoline as GABA-A Inhibitor with Anticonvulsant Profile: An *in-silico* approach. *American Journal of Pharmacological Sciences*, 1, 116-20. Accessed November 02, 2019.doi: 10.12691/ajps-1-6-2.

(b) Amir, Mohd., S. Asif., Ali, I. and Hassan, M. Z. (2012). Synthesis of benzothiazole derivatives having acetamido and carbothio-amidopharmacophore as anticonvulsant agents. *Medicinal Chemistry Research*, 21, 2661–70. Accessed November 02, 2019. doi: 10.1007/s00044-011-9791-1.

(c) Navale, A., Pawar, S., Deodhar, M. N. and Kale, A., (2013). Synthesis of substituted benzo[*d*]thiazol-2-ylcarbamates as potential anticonvulsants. *Medicinal Chemistry Research*, 22, 4316–4321. Accessed November 02, 2019. doi: 10.1007/s00044-012-0434-y

(d)Patrick, J., François, A., Michel, B., Jean, C. B., Alain, B., Yvette, B., Marie, A. C., Doble, A., Gilles, D., Claudine, D. H., Marie, H. D., Jean, M. D., Pierre, G., Claude, G., Eliane, H., Bernard, J., Roselyne, K., Sylvie, G., Philippe, H., Pierre, M. L., Joseph, L. B., Mireille, M., Jean, M. M., Conception, N., Martine, P., Odile, P., Jeremy, P., Jean, R., Michel, R., Jean, M S. and Serge, M. (1994). Riluzole series. Synthesis and *in vivo* "antiglutamate" activity of 6- substituted- 2-benzothiazolamines and 3-substituted-2-imino-benzothiazolines. *Journal of Medicinal Chemistry*, 42, 2828–43. Accessed November 02, 2019. doi: 10.1021/jm980202u.

(e) Rana, A., Siddiqui, Nadeem., Khan, Suroor, A., Haque, S. E. and Bhat, M. A.. (2008). *N*-[(6-Substituted-1,3-benzothiazole-2-yl)amino]carbonothioyl-2/4-substituted benzamides: synthesis and pharmacological evaluation. *European Journal of Medicinal Chemistry*, 43, 1114–22. Accessed November 02, 2019. doi: 10.1016/j.ejmech.2007.07.008.

[62] Hassan, M. Z., Khan, S. A. and Amir, M. (2012). Design, synthesis and evaluation of *N*-(substituted benzothiazol-2-yl)amides as anticonvulsant and neuroprotective. *European Journal of Medicinal Chemistry*, 58, 206–13. Accessed November 02, 2019. doi: 10.1016/j.ejmech.2012.10.002.

[63] Ugale, V. G., Patel, H. M., Wadodkar, S. G., Bari, S. B., Shirkhedkar, A. A. and Surana, S. J. (2012). Quinazolino-benzothiazoles: fused pharmacophores as anticonvulsant agents. *European Journal of Medicinal Chemistry*, 53, 107–13. Accessed November 04, 2019. doi: 10.1016/j.ejmech.2012.03.045.

[64] Mariappan, G., Prabhat, P., Sutharson, L., Banerjee, J., Patangia, U. and Nath, S. (2012). Synthesis and antidiabetic evaluation of benzothiazole derivatives. *Journal of the Korean Chemical Society*, 56, 251-56. Accessed November 03, 2019. doi: 10.5012/jkcs.2012.56.2.251.

[65] Patil, V. S.., Nandre, K. P., Ghosh, S., Rao, V. J., Chopade, B. A., Sridhar, B., Bhosale, S. V. and Bhosale, S. V. (2013). Synthesis, crystal structure and antidiabetic activity of substituted (E)-3-(Benzo [d] thiazol-2-ylamino) phenylprop-2-en-1-one. *European Journal of Medicinal Chemistry*, 59, 304–09. Accessed November 03, 2019. doi: 10.1016/j.ejmech.2012.11.020.

[66] Jeon, R., Kim, Y. J., Cheon, Y. and Ryu, J. H. (2016). Synthesis and biological activity of [[(heterocycloamino)alkoxy]benzyl]-2,4-thiazolidinediones as PPART agonists." *Archives of Pharmacal Research*, 29, 394-99. Accessed November 03, 2019. doi: 10.1007/BF02968589.

[67] Munirajasekhar, D., Himaj, M., Mali, S. V., Karigar, A. and Sikarwar, M. (2011). Synthesis and anthelmintic activity of 2-amino-6-

substituted benzothiazoles. *Journal of Pharmacy Research*, 2, 114–17. Accessed November 03, 2019.
url: http://jprsolutions.info/newfiles/journal-file-56e961eab0db86.31965620.pdf.

[68] Suresh, C. H., Rao, J. V., Jayaveera, K. N. and Subudhi, S. K. (2011). Synthesis and anthelmintic activity of 3-(2-hydrozino benzothiazole) substituted indole-2-one *International Journal of Pharmaceutics*, 2, 257–61. Accessed November 03, 2019.
url:https://pdfs.semanticscholar.org/66d2/953bafb906adc0c5989c8eb9d91ff58da190.pdf.

[69] Ongarora, D. S. B., Gut, J., Rosenthal, P. J., Masimirembwa, C. M. and Chibale, K. (2012). Benzoheterocyclicamodiaquine analogues with potent antiplasmodial activity: synthesis and pharmacological evaluation. *Bioorganic & Medicinal Chemistry Letters*, 22, 5046–50. Accessed November 05, 2019. doi: 10.1016/j.bmcl.2012.06.010.

[70] Venugopala, K. N., Krishnappa, M., Nayak, S. K., Bhat, K. S., Jayashankaragowda, P. V., Raju, K. C., Raquel, M. G. and Bhart, O. (2013). Synthesis and antimosquito properties of 2,6-substituted benzo[d]thiazole and 2,4-substituted benzo[d]thiazole analogues against Anopheles arabiensis. *European Journal of Medicinal Chemistry*, 65, 295–303.Accessed November 05, 2019. doi:10.1016/j.ejmech.2013.04.061.

[71] Gilani, S. J., Nagarajan, K., Dixit, S. P., Taleuzzaman, M. and Khan, S. A. (2016). Benzothiazole incorporated thiazolidin-4-ones and azetidin-2-ones derivatives: Synthesis and *in vitro* antimicrobial evaluation. *Arabian Journal of Chemistry*, 9, 1523-31. Accessed November 05, 2019. doi: 10.1016/j.arabjc.2012.04.004.

[72] Singh, M., Singh, S. K., Gangwar, M., Nath, G. and Singh, S. K. (2014). Design, synthesis and mode of action of some benzothiazole derivatives bearing an amide moiety as antibacterial agents. *Royal Society of Chemistry*, 4, 19013–23. Accessed November 05, 2019. doi: 10.1039/C4RA02649G

[73] Tomi, I. H. R.., Tomma, J. H., Ali, H. R. A. and Dujaili, A. H. (2015). Synthesis, characterization and comparative study the microbial

activity of some heterocyclic compounds containing oxazole and benzothiazole moieties. *Journal of Saudi Chemical Society*, 19, 392-98.Accessed November 05, 2019.
doi: 10.1016/j.jscs.2012.04.010.

[74] Singh, M. K., Tilak, R., Nath, G., Awasthi, S. K. and Agarwal, A. (2013). Design, synthesis and antimicrobial activity of novel benzothiazole analogs. *European Journal of Medicinal Chemistry*, 63, 635–44.Accessed November 07, 2019. doi: 10.1016/j.ejmech.2013.02.027.

[75] Maddila, S., Gorle, S., Seshadri, N., Lavanya, P. and Jonnalagadda, S. B. (2016). Synthesis, antibacterial and antifungal activity of novel benzothiazole pyrimidine derivatives. *Arabian journal of Chemistry*, 9, 681-87.Accessed November 07, 2019.
doi: 10.1016/j.arabjc.2013.04.003.

[76] Mahtab., R., Srivastava, A., Gupta, N., Kushwaha, S. K. and Tripathi, A. (2014). Synthesis of novel 2-benzylbenzo[d] thiazole-6-sulfonamide derivatives as potential antiinflammatory agent. *Journal of Chemical and Pharmaceutical Sciences*, 7, 34-38. Accessed November 07, 2019. url:https://pdfs.semanticscholar.org/6231/4980169fddfd387a2af22a33bea8c2f7b687.pdf?_ga=2.105144323.1648233827.1576777791-360881681.1576777791.

[77] Shafi, S., Alam, M. M., Mulakayala, N., Mulakayala, C., Arunasree, V, G., Kalle, M.., Pallu, R. and Alam. M. S. (2012). Synthesis of novel 2-mercapto benzothiazole and 1,2,3-Triazole based Bis-heterocycles: Their anti-inflammatory and anti-nociceptive activities. *European Journal of Medicinal Chemistry*, 49, 324-33. Accessed November 07, 2019. doi: 10.1016/j.ejmech.2012.01.032.

[78] Kumar, A. K. V. and Gopalakrishna, B. (2014). Synthesis and Biological, Pharmacolgical Activites of Bioactive Benzothiazole Deravatives. *Research and Reviews: Journal of Pharmacy and Pharmaceutical Sciences*, 3, 50-54.Accessed November 07, 2019. url:http://www.rroij.com/open-access/synthesis-and-biological-pharmacolgical-activites-of-bioactive-benzothiazole-deravatives-50-54.pdf.

[79] Velingkar, V. S., Ahire, D. C., Kolhe, N. S., Shidore, M. M. and Pokharna, G. (2011). Synthesis, characterization, biological evaluation and ADME studies using *in silico* techniques of novel derivatives of benzothiazolyl-amides as non-acidic anti-inflammatory agents. *International Journal of Pharmaceutical Sciences and Research*, 2, 183-88. Accessed November 07, 2019. url:http://ijpsr.com/bft-article/synthesis-characterization-biological-evaluation-and-adme-studies-using-in-silico-techniques-of-novel-derivatives-of-benzothiazolyl-amides-as-non-acidic-anti-inflammatory-agents/?view=fulltext.

[80] Verma, B. K., Martin, A. and Singh. A. K. (2014). Synthesis, Characterization and evaluation of Anti-inflammatory and Analgesicactivity of Benzothiazole derivatives. *Indian Journal of Pharmaceutical and Biological Research*, 2, 84-89. Accessed November 07, 2019. file:///C:/Users/DR/Downloads/Synthesis_Characterization_and_evaluation_of_Anti-.pdf.

[81] Kaur, H., Kumar, S., Singh, I., Saxena, K. K. and Kumar, A. (2010). Synthesis, characterization and biological activity of various substituted benzothiazole derivatives. *Digest Journal of Nanomaterials and Biostructures*, 5, 67-76. Accessed November 07, 2019. https://pdfs.semanticscholar.org/ab4a/6990b87091499d15683280404 5c117955595.pdf.

[82] Pareek, D., Chaudhary, M., Pareek, P. K., Kant, R., Ojha, K.G., Pareek, R., Iraqia, S. M. U. and Pareeka, A. (2010). Synthesis of some bioactive 4-thiazolidinone derivatives incorporating benzothiazolemoiety. *Der Chemica Sinica*, 1, 22-35. Accessed November 07, 2019. https://www.econ-jobs.com/research/16005-Synthesis-of-some-bioactive-4-thiazolidinone-derivatives-incorporating-benzothiazole-moiety.pdf.

[83] Mahran, M. A., El-Nassry, S. M. F., Allam. S. R. and El-Zawawy. L. A. (2003). Synthesis of some new benzothiazole derivatives as potential antimicrobial and antiparasitic agents. *Pharmazie*, 58, 527–30. Accessed November 07, 2019. doi: 10.1002/chin.200346139.

[84] Arora, P., Das, S., Mahendra, S. Ranawat., Arora, N. and Gupta, M. M. (2010). Synthesis and Biological Evaluation of Some Novel Chromene-2-one Derivatives for Antipsychotic Activity. *Journal of Chemical and Pharmaceutical Research*, 2(4), 317-23. Accessed November 08, 2019. http://www.jocpr.com/articles/synthesis-and-biological-evaluation-of-some-novel-chromene2one-derivatives-for-antipsychotic-activity.pdf.

[85] Chaudhary, M., Pareek, D., Pawan, K. Pareek., Kant, R., Krishan, G. Ojha. and Pareek, A.(2010). Synthesis of some biologically active benzothiazole derivatives. Scholars Research Library, *Der Pharma Chemica*, 2(5), 281-93. Accessed November 10, 2019. https://www.econ-jobs.com/research/16005-Synthesis-of-some-biologically-active-benzothiazole-derivatives.pdf.

[86] Yar, M. S. and Ansari. Z. H. (2009). Synthesis and *in vivo* diuretic activity of biphenyl benzothiazole-2-carboxamidederivatives. *Acta Poloniae Pharmaceutica Drug Research*, 66, 387-92. Accessed November 10, 2019.
https://www.ptfarm.pl/pub/File/Acta_Poloniae/2009/4/387.pdf.

[87] Verma, S. M., Dadheech, M. and Ram, P. M. (2012). Design and Synthesis of some Benzothiazole Analogs as α2A Receptor Antagonist. *Journal of Pharmaceutical Science and Technology*, 1(2), 30-35. Accessed November 10, 2019. https://pdfs.semanticscholar.org/18fe/6f6adcc0ed8acdae151433908433f2a69dd7.pdf?_ga=2.251627462.1728529531.1576755779-2033189515.1576755779.

[88] Mahajan, D. F., Bhosale, J. D. and Bendre, R. S.. (2013). Synthesis, Characterization and Plant Growth Regulator Activity of Some Substituted 2-Amino Benzothiazole Derivatives. *Journal of Applicable Chemistry*, 2, 765-71. Accessed November 10, 2019. https://pdfs.semanticscholar.org/7653/bf236bbcf661b1cdea4f262e288c40c9a749.pdf?_ga=2.258270794.1728529531.1576755779-2033189515.1576755779.

[89] (a) Liao, C., Kim, U. and Kannan, K. (2015). A Review of Environmental Occurrence, Fate, Exposure, and Toxicity of

Benzothiazoles. *Environmental Science & Technology*, 52, 5007-26. Accessed November 10, 2019. doi: 10.1021/acs.est.7b05493.

(b) Record of CAS RN 21564-17-0 in the GESTIS Substance Database of the Institute for Occupational Safety and Health. Accessed November 10, 2019.

(c) Engin, Bagda. (2000). *Biocides in Building Coatings.* Expert publisher. (Engin Bagda 2000, 59). ISBN 3-81691861-1.

(d) Levant. & Ezra. (2014). *Groundswell: The Case for Fracking.* McClelland & Stewart. Levant and Ezra 2014, 192).

[90] Hornung, M. W., Kosian, P. A., Haselman, J. T., Korte, J. J., Challis, K., Macherla, C., Nevalainen, E. and Degitz. S. J. (2015). *In vitro, ex vivo* and *in vivo* determination of thyroid hormone modulating activity of benzothiazoles. *Toxicoligical Sciences*, 146, 254-64. Accessed November 10, 2019. doi: 10.1093/toxsci/kfv090.

[91] Yan, Y., Jiang, W., Li, N., Ma, M., Rao, K. and Wang, Z.(2014). Application of the SOS/umu test and high content *in vitro* micronucleus test to determine genotoxicity and cytotoxicity of nine benzothiazoles. *Journal of applied Toxicology*, 34, 1400-08. Accessed November 10, 2019. doi: 10.1002/jat.2972.

[92] Sheikh, J. H., Aoshima, H., Koda, H. and Kiso, Y.(2003). Effects of coffee components on the response of GABA(A) receptors expressed in Xenopus oocytes. *Journal of Agricultural Food Chemistry*, 51(26), 7568-75. Accessed November 10, 2019.doi: 10.1021/jf0303971.

[93] (a) Bogert, M. T. and Husted, H. G. (1932). Contribution to the pharmacology of the benzothiazoles. *Journal of Pharmacology and Experimental Therapeutics*, 45, 189-207. Accessed November 10, 2019. http://jpet.aspetjournals.org/content/45/2/189.

(b) Ikarashi, Y., Tsuchiya, T. and Nakamura, A. (1993). Evaluation of contact sensitivity of rubber chemicals using the murine local lymph node assay. *Contact Dermatitis*, 28, 77-80. Accessed November 10, 2019. doi:10.1111/j.1600-0536.1993.tb03345.x.

[94] Bustnes, O., Datta G. N., Koomey, J. E. and Lovins, A. (2007). Asphalt, feedstocks and lubricants. "Technical annex in Winning the oil endgame (Bustnes and Lovins Chapter 14. 2007). Accessed

November 10, 2019. http://www.oilendgame.com/pdfs/ TechAnnx/TechAnnx14.pdf.

[95] Seo, K. W., Park, M., Kim, J. G., Kim. T. W. and Kim. H. J. (2000). Effects of benzothiazole on the xenobiotic metabolizing enzymes and metabolism of acetaminophen. *Journal of Applied Toxicology*, 20, 427-30. Accessed November 10, 2019. doi:10.1002/1099-1263 (200011/12)20:63.0.CO;2-# ·

[96] Korhonen, A., Hemminki, K. and Vainio, H. (1983). Toxicity of rubber chemicals towards three-day chicken embryos. *Scandinavian Journal of work, Environment & Health.*, 9(2), 115-119. Accessed November 10, 2019. doi: 10.5271/sjweh.2435

[97] Whittaker, M. H., Gebhart, A. M., Miller, T. C. and Hammer, F. (2004). Human health risk assessment of 2-mercaptobenzothiazole in drinking water. *Toxicology and Industrial Health*, 20(6-10), 149-63. Accessed November 10, 2019. doi:10.1191/0748233704th199oa.

[98] Kogevinas, M., Sala, M., Boffetta, P., Kazerouni, N., Kromhout, H. and Hoar-Zahm, S. (1998). Cancer risk in the rubber industry: a review of the recent epidemiological evidence. *Occupational Environmental Medicine*, 55(1), 1-12. Accessed November 10, 2019. doi: 10.1136/oem.55.1.1.

BIOGRAPHICAL SKETCH

Ahmed Kamal

Affiliation: Pro-Vice Chancellor, Jamia Hamdard, New Delhi 110062, India and Professor, Pharmaceutical Chemistry, School of Pharmaceutical Education and Research (SPER), Jamia Hamdard

Formerly Outstanding Scientist & Head, Medicinal Chemistry and Pharmacology, CSIR - Indian Institute of Chemical Technology (IICT), Hyderabad, India & Project Director, National Institute of Pharmaceutical Education and Research (NIPER), Hyderabad

Education:

- Post-Doc. Research (Medicinal Chemistry), University of Portsmouth, UK, 1988-89
- PhD (Chemistry), Aligarh Muslim University, Aligarh, 1982 (worked at IICT)
- MPhil (Chemistry), Aligarh Muslim University, Aligarh, 1979 (worked at IICT)
- MSc (Organic Chemistry), Aligarh Muslim University, Aligarh, 1977
- BSc (Chemistry, Biology), Osmania University, Hyderabad, 1975

Present Research Interests: Multi-disciplinary research programmes including organic synthesis, medicinal, combinatorial and green chemistry including chemical biology and biocatalysis

Design and synthesis of gene-targeting compounds as new and novel anticancer agents, and their targeted delivery as prodrugs.

Significant Research Contributions:

Development of Anticancer Therapeutics: The discovery of potent, selective and less toxic anticancer agents has been considered as one of the major challenges in medicinal chemistry. Significant efforts have been made to design and synthesize a large number of heterocyclic hybrids and their conjugates wherein at least two biologically well-established components were brought in to a single moiety that could interact or sometimes enhance the biological effect for the same target. In this pursuit, structural modifications on the pyrrolo[2,1-*c*]benzodiazepine (PBD) ring system has been explored extensively. These results provided further inputs to explore the combination of certain non-covalent interacting groups with a PBD moiety that led to the design and synthesis of a variety of hybrids and conjugates. Most of the anticancer agents using chemotherapy of cancer lack selectivity towards tumour cells leading to secure side effects and dose limitation. In this context, the development of glycoside prodrugs of PBDs has been investigated. This provided improved selectivity of the PBDs towards cancer tissues, through β-galactosidase based ADEPT and PMT strategies.

Moreover, a large number of DNA topoisomerase II and tubulin polymerization inhibitors as well as inducers of apoptosis have been designed, synthesized and evaluated.

Development of New Antitubercular Agents: Many heterocyclic scaffolds like phthalamido/naphthalimido linked phenazines, 1,2,4-benzothiadiazines, benzothiazole conjugates, arylsulfonamido-oxazolidinones and thialactone based conjugates have been designed and investigated to evaluate their antitubercular potential.

Biocatalysis/Biotransformations and Biofuels: A large number of enantiomerically pure chiral intermediates have been obtained by lipase catalysed transesterification processes.

Career Profile Including Managerial Role: The research and development activities during the career represent many conceptual and original ideas with experimental excellence that is in tune with the priorities and requirements. The contributions made are mostly relating to multidisciplinary research programmes, which include Organic Synthesis, Medicinal Chemistry, Combinatorial Chemistry, Green Chemistry and Chemical Biology. Played an important role in the formulation and development of a large number of projects/programmes with industry collaboration/sponsorship that led to fruitful industry-institute linkages.

Establishment of NIPER, Hyderabad/BTIC: As a Project Director of National Institute of Pharmaceutical Education and Research (NIPER), Hyderabad for about seven years, has taken up several measures for the overall growth and improvement of this institute of national importance. Many new concepts have been introduced with respect to a large number of aspects, such as administrative, academics/examinations and research activities. New disciplines that could be important for this region including

Ph.D Programmes as well as management courses in pharmaceutical sciences have been introduced. Significant inputs have been provided for the conceptualization of the National Centre for Research and Development in Bulk Drugs (NCRDBD).

Played an important role in the conceptualization and establishment of Biotechnology Incubation Centre (BTIC) at the Biotech Park in the Genome Valley. This facility has attracted several entrepreneurs to setup their own biotech companies in and around Hyderabad.

Impact of Research Work on Industry/Society/Environment: A large number of projects were undertaken in collaboration or by sponsorship with industry for the development of new chemical entities and innovative process technologies. The research work relating to new chemical entities has been focused towards the affordable healthcare (cancer therapeutics). Substantial number of process technologies have been investigated in collaboration/sponsorship with industry, particularly, in the development of environmentally benign processes by utilizing microbes/enzymes as biocatalysts with an endevour to bring organic synthetic processes closure to that practised by nature. This challenging area of research has considerable impact on the environment apart from the development of cost-effective processes.

Research programmes in association with industry were conceptualised. Some of the industries that were associated with are: Yamanuchi Pharmaceuticals Limited (Japan), Mitsubishi (Japan), Marubeni (Japan), SheratoriPharma (Japan), Dupont (USA), Spirogen Pharmaceuticals (UK), Evolva (Switzerland), Ranbaxy Laboratories (New Delhi), Lupin Laboratories (Pune), Acoris, (Pune) and Pedilite (Mumbai).

Several academic collaborations were developed internationally and nationally and some of these institutes are - Imperial College London, Kings College, London, University of Wuppertal, Germany, University of Greifswald, Germany, University of Cape Town, University of Southern California, USA, ACTREC, Mumbai, CCMB, Hyderabad and IISc, Bangalore.

Indo-US Clean Energy Research Initiative: Coordinated the US-India Consortium for the Development of Sustainable Advanced Lignocellulosic Biofuel Systems under the Second Generation Biofuels.

UK-India Education and Research Initiative (UKIERI): In the first programme, research work on the Biomedical Solutions between India and UK was carried out through the Imperial College, London. Then in the second programme, the Institute of Pharmaceutical Sciences, King's College London and the Indian team have coinvented a new nanoparticle platform that may have substantial implication for new nanomedicine therapeutics.

Professional Appointments:

Oct 3, 2017 till date	Pro-Vice Chancellor, JamiaHamdard, New Delhi, India
June 9, 2015 to May 2016	Outstanding Scientist (Director level) CSIR - Indian Institute of Chemical Technology, Hyderabad, India
Sept. 2009 to April 2016	Project Director, National Institute of Pharmaceutical Education and Research [NIPER], Hyderabad
April 1, 2015 to June 8, 2015	Acting Director CSIR - Indian Institute of Chemical Technology, Hyderabad, India
April 8, 2013 - March 31, 2015	Outstanding Scientist (Director level) CSIR - Indian Institute of Chemical Technology, Hyderabad, India
Sept. 2012- April 7, 2013	Acting Director CSIR - Indian Institute of Chemical Technology, Hyderabad, India
2010-2012	Outstanding Scientist (Director level) CSIR - Indian Institute of Chemical Technology, Hyderabad, India

2007 - 2010	Chief Scientist (Scientist-G), Indian Institute of Chemical Technology, Hyderabad, India
1992 - 2007	Scientist at different levels, Indian Institute of Chemical Technology, Hyderabad, India
1993 - 94	Visiting Scientist, University of Alberta, Edmonton, Canada
1983 - 92	Scientist-B and Scientist-C, Indian Institute of Chemical Technology, Hyderabad, India
1977-82	CSIR -Junior/Senior Research Fellow

Honours:

- YMSA Young Scientist Award from MAAS & TWAS - 1988
- CSIR Young Scientist Award in Chemical Sciences - 1991
- Fellow of National Academy of Sciences, India - 1999
- Best Patent Award from the Indian Drug Manufacturers Association (IDMA) - 2005
- Medal from the Chemical Research Society of India (CRSI) for contributions to research in Chemistry - 2005
- Ranbaxy Research Award in the field of Pharmaceutical Sciences - 2005
- UKIERI Standard Award for Biomedical Solutions between India and UK - 2006
- Andhra Pradesh Scientist Award in Chemical Sciences by A P State Council of Science & Technology -2007
- OPPI Scientist Award from the Organization of Pharmaceutical Producers of India - 2009
- Fellow of Andhra Pradesh Academy of Sciences (FAPSc) - 2010
- Fellow of Royal Society of Chemistry (FRSC) - 2011
- Most Outstanding Researcher in the field of Chemistry by Careers 360 -2018

Publications from the Last 3 Years:

[1] Jadala, C., Sathish, M., Anchi, P., Tokala, R., Lakshmi, U. J., Ganga, R., Shankaraiah, N., Godugu, C. & Kamal, A. (2019). Synthesis of combretastatin-A4 carboxamides that mimic sulfonyl piperazines by a molecular hybridization approach: *in vitro* cytotoxicity evaluation and inhabitation of tubulin polymerization. *Chem Med Chem.*, 14, 1.

[2] Jadala, C., Prasad, B., Prashant, A. V. G., Shankaraiah, N. & Kamal, A. (2019). Transition metal-free one-pot synthesis of substituted pyrroles by employing aza-witing reaction. *RSC Adv.*, 2019, (in press).

[3] Prasad, B., Phanindrudu, M., Tiwari, D. K. & Kamal, A. (2019). Transition metal free one-pot tandem synthesis of 3-Ketoisoquinolines from aldehydes and phenacyl azides. *J. Org. Chem.*, (in press).

[4] Mani, G. S., Donthiboina, K., Shankaraiah, N. & Kamal, A. (2019). Iodine-promoted one-pot synthesis of 1,3,4-Oxadiazole scaffolds via sp^3 C-H functionalization of azaarenes. *New J. Chem.*, 43, 15999.

[5] Mani, G. S., Donthiboina, K. S., Shaik, S. P., Shankaraiah, N. & Kamal, A. (2019). Iodine-mediated C–N and N–N bond formation: a facile one-pot synthetic approach to 1,2,3- triazoles under metal-free and azide-free conditions. *RSC Adv.*, 9, 27021.

[6] Shareef, M. A., Sirisha, K., Sayeed, I. B., Khan, I., Ganapathi, T., Akbar, S., Kumar, C. G., Kamal, A. &Babu, B. N. (2019).Synthesis of new triazole fused imidazo[2,1-b]thiazole hybrids with emphasis on staphylococcus aureus virulence factors. *Bioorg. Med. Chem. Lett.*, (in press).

[7] Paidakula, S., Nerella, S., Vadde, R., Kamal, A. & Kankala, S. (2019). Design and synthesis of 4β-Acetamidobenzofuranone-podophyl-lotoxin hybrids and their anti-cancer evaluation. *Bioorg. Med. Chem. Lett.*, 29, 2153.

[8] Haider, K., Rahaman, S., Yar, M. S. & Kamal, A. (2019). Tubulin inhibitors as novel anticancer agents: an overview on patents (2013-2018). *Expert Opinion on Therapeutic Patents*, 29, 623.

[9] Nainwal, L. M., Alam, M. M., Shaquiquzzaman, M., Marella, A. & Kamal, A. (2019). Combretastatin based compounds with therapeutic

characteristics: a patent review. *Expert Opinion on Therapeutic Patents*,2019, *29*,703.

[10] Reddy, V. G., Reddy, T. S., Jadala, C., Reddy, M. S., Sultana, F., Akunri, R., Bhargava, S. K., Wlodkowic, D., Srihari, P. & Kamal, A. (2019). Pyrazolo-benzothiazole hybrids: Synthesis and evaluation of anticancer properties with VEGFR-2 Kinase inhibition and anti-angiogenesis in transgenic zebrafish *in vivo* model. *Eur. J. Med. Chem.*, (in press).

[11] Berishvili, V. P., Perkin, V. O., Voronkov, A. E., Radchenko, E. V., Riyaz, S., Reddy, V. R. C., Pillay, V., Kumar, P., Choonara, Y. E., Kamal, A. & Palyulin, V. A. (2019). Time-domain analysis of molecular dynamics trajectories using deep neural networks: application to activity ranking of tankyrase inhibitors. *J.Chem. Inf. Model*, *59*, 3519.

[12] Radhakrishnan, S., Syed, R., Takei, H., Kobayashi, I. S., Nakamura, E., Sultana, F., Kamal, A. & Tenen, D. G. (2019). Styryl Quinazolinones and its ethynyl derivatives induce myeloid differentiation.Kobayashi, S. S., *Bioorg. & Med. Chem.*, *29*, 2286.

[13] Jadala, C., Sathish M., Reddy, T. S., Tokala, R., Bhargava, S. K., Shankaraiah, N., Nagesh, N. & Kamal, A. (2019). Synthesis and *in vitro* cytotoxicity evaluation of β-carboline-combretastatin carboxamides as apoptosis inducing agents: DNA intercalation and topoisomerase-II inhibition. *Bioorg. & Med. Chem.*, *27*, 3285.

[14] Sultana, F., Saifi, M. A., Syed R., Mani, G. S., Shaik, S. P., Osas, E. G., Godugu, C., Shahjahan, S. & Kamal, A. (2019).Synthesis of 2-anilinopyridyl linked benzothiazole hydrazones as apoptosis including cytotoxic agents *New J. Chem.*, *43*, 7150.

[15] Sunkari, S., Bonam, S. R., Rao, A. V. S., Riyaz, S., Nayak, V. L., Kumar, H. M. S., Kamal, A. &Babu, B. N. (2019). Synthesis and biological evaluation of new bisindole-imidazopyridine hybirds as apotosis inducers.*Bioorganic Chemistry*, (in press).

[16] Rahim A., Syed, R., Poornachandra, Y., Malik, M. S., Reddy, Ch. R., Alvala, M., Boppana, K., Sridhar, B., Amancy, R. & Kamal, A. (2019). Synthesis and biological evaluation of phenyl-amino-pyrimidne and

indole/oxindole conjugates as potential BCR-ABL inhibitors. *Medicinal Chemistry Research*, 28, 633.

[17] Shaik, S. P., Reddy, T. S., Sunkari, S., Rao, A. V. S., Babu, K. S., Bhargava, S. K. &Kamal, A. (2019). Synthesis of benzo[d]imidazo[2,1-b]thiazole-propenone conjugates as cytotoxic and apoptotic inducing agents. *Anti-Cancer Agents in Medicinal Chemistry*, 19.

[18] Khan, I., Ganapathi, T., Shareef, M. A., Shaik, A. B., Akbar, S., Ranjanna, A., Kamal, A. & Kumar, C. G. (2019). One pot synthesis and biological evalution of arylopropenone aminochalcone conjugates as potential apoptic inducers. *Chemistry Select*, 4, 4672.

[19] Shareef, M. A., Sirisha, K., Khan, I., Sayeed, I. B., Gopathi, R., Kumar, C. G., Kamal, A. & Bathini, N. B. (2019). Design, synthesis and antimicrobial evaluation of 1,4-dihydroindeno[1,2-c]pyrazole tethered carbohydrazide hybrids: Exploring their in silico ADMET, antibiofilm, ergosterol inhibition and ROS inducing potential. *Med. Chem. Commun.*, (in press).

[20] Khan, I., Sirisha, K., Shareef, M. A., Ganapathi, T., Shaik, A. B., Shekar, K. C., Kamal, A. & Kumar, C. G. (2019).Synthesis of new bis-pyrazole linked hydrazides and their *in vitro* evaluation as antimicrobial and anti-biofilm agents: A mechanistic role on ergosterol biosynthesis inhibition in *Candida albicans*. *Chem.Biol. & Drug Design*, 94, 1339.

[21] Nagaraju, B., Kovvuri, J., Kumar, C. G., Routhu, S. R., Shareef, M. A., Kadagathur, M., Adiyala, P. R., Alavala, S., Nagesh, N. & Kamal, A. (2019). Synthesis and biological evaluation of pyrazole linked benzothiazole-β-naphthol derivatives as topoisomerase I inhibitors with DNA binding ability. *Bioorg. & Med. Chem.*, 27, 708.

[22] Routhu, S. R, Chary, R. N., Shaik, A. B., Prabhakar, S., Kumar, C. G. & Kamal, A. (2019). Induction of apoptosis in lung carcinoma cells by antiproliferative cyclic lipopeptides from marine algicolous isolate bacillus atrophaeus strain AKLSR1.*Process Biochemistry*, 79, 142.

[23] Shareef, M. A., Rajpurohit, H., Sirisha, K., Sayeed, I. B., Khan, I., Kadagathur, M., Ganapathi, T., Kumar, C. G., Kamal, A. &Babu, B.

N. (2019). Design, synthesis and biological evaluation of substituted (1-(4-chlorobenzyl)-1H-indol-3-yl) 1H-(1,2,3-triazol-4-yl) methanones as antifungal agents. *Chemistry Select*, *4*, 2258.

[24] Khan, I., Garikapati, K. R., Setti, A., Shaik, A. B., Makani, V. K. K., Shareef, M. A., Rajpurohit, H., Vangara, N., Pal-Bhadra., M., Kamal, A. & Kumar, C. G. (2019). Design, synthesis, *in silico* pharmacokinetics prediction and biological evaluation of 1,4-dihydroindeno[1,2-c]pyrazole chalcone as EGFR/Akt pathway inhibitors. *Eur. J. Med. Chem.*, *163*, 636.

[25] Donthiboina, K., Anchi, P., Ramya, P. V. S., Karri, S., Srinivasulu, G., Godugu, C., Shankaraiah, N. & Kamal, A. (2019). Synthesis of substituted biphenyl methylene indolinones as apoptosis inducers and tublin polymerization inhibitors. *Bioorg. Chem.*, *86*, 210.

[26] Malik, M. S., Seddigi, Z. S., Bajee, S., Azeeza, S., Riyaz, S., Ahmed, S. A., Althagafi, I. I., Jamal, Q. M. S. & Kamal, A. (2019). Multicomponent access to novel proline/cyclized cysteine tethered monastrol conjugates as potential anticaner agents. *J. Saudi Chem. Soc.*, *23*, 503.

[27] Aaghaz, S., Gohel, V. & Kamal, A. (2019). Peptides as potential anticancer agents. *Curr. Top. Med. Chem.*, (in press).

[28] Prasad, B., Nayak, V. L., Srikanth, P. S., Baig, M. F., Subba Reddy, N. V., Babu, K. S. & Kamal, A. (2019). Synthesis and biological evaluation of 1-benzyl-N-(2-(phenylamino)pyridin-3-yl)-1H-1,2,3-triazole-4-carboxamides as antimitotic agents. *Bioorg. Chem.*, *83*, 535.

[29] Kamal, A., Shankaraiah, N., Regur, P., Gaddam, K., Tokala, R., Sunkari, S., Nayak, V. L., Nagesh, N. & Rao, N. S. (2019). Design and synthesis of DNA-intercalative naphthalimide-benzothiazole/ cinnamide derivates: Cytotoxicity evaluation and topoisomerase-IIα inhibition. *Med. Chem. Comm.*, *10*, 72.

[30] Sultana, F., Manasa, K. L., Shaik, S. P., Bonam, S. R. & Kamal, A. (2018). Zinc Dependent Histone Deacetylase Inhibitors in Cancer Therapeutics: Recent Update. *Curr. Med. Chem.*, *25*, 1.

[31] Tangella, Y., Manasa, K. L., Harikrishna, N., Sridhar B., Kamal, A. &Babu, B. N. (2018). Regioselective Ring Expansion of Isatins with

In Situ Generated α-Aryldiazomethanes: Director Access to Viridicatin Alkaloids. *Org. Lett.*, 2018, *20*, 3639.

[32] Kumar, N. P., Kumari, S. S., Lakshmi, U. J., Tokala, R., Shankaraiah, N. & Kamal, A. (2018). Sulfamic acid promoted one-pot synthesis of phenanthrene fuseddihydrodibenzo-quinolinones: Anticancer activity, tubulin polymerization inhibition and apoptosis inducing studies. *Bioorg. Med. Chem.*, *26*, 1996.

[33] Sayeed, I. B., Vishnuvardhana, M. V. P. S., Nagarajan, A., Kantevari, S. & Kamal, A. (2018). Imidazopyridine linked triazoles as tubulin inhibitors, effectively triggering apoptosis in lung cancer cell line. *Bioorg. Chem*, *80*, 714.

[34] Kumara N. P., Vanjaria, Y., Thatikondab, S., Pooladandab V., Sharma P., Sridhard B., Godugub, C., Kamal, A. & Nagula S. (2018). Synthesis of enamino-2-oxindoles via conjugate addition between α-azido ketones and 3-alkenyl oxindoles: Cytotoxicity evaluation and apoptosis inducing studies.*Bioorg. Med. Chem. Lett.*, *28*, 3564.

[35] Thatikonda, S., Tangella, Y., Krishna, H., Nekkanti, S., Dushantrao, S. C., Cherukommu, S., Srinivas, G., Tokala, R., Nagesh, N., Manda, S., Nagula, S. & Kamal, A. (2018). Synthesis of DNA interactive C3-trans-cinnamide linked β-carboline conjugates as potential cytotoxic and DNA topoisomerase I inhibitors. *Bioorg. Med. Chem. Lett.*, *26*, 4916.

[36] Shankaraiah, N., Namballa, H. R., Sultana, F., Sunkari, S., Tangella, Y., Subba, R. A. V., Mani, S. G. & Kamal, A. (2018). Molecular idone-catalysed oxidative CO-C(alkyl) bond cleavage of aryl/heteroaryl alkyl ketones: An efficient strategy to access fused polyheterocycles. *New J. Chem.*, *42*, 15820.

[37] Digwal, C. S., Yadav, U., Ramya, P. V. S., Swain, B. & Kamal, A. (2018). Vanadium-catalyzed *N*-benzoylation of 2-aminopyridinesvia oxidative C(CO)–C(CO) bond cleavage of 1,2-diketones,*N*→*N'* aroyl migration and hydrolysis of 2-(diaroylamino) pyridines. *AsianJ. Org. Chem.*, *7*, 865.

[38] Ramya, P. V. S., Guntuku, L., Angapelly, S., Digwal, C. S., Lakshmi, U. J., Sigalapalli, D. K. J., Babu, B. N., Naidu, V. G. M. & Kamal, A.

(2018). Synthesis and biological evaluation of curcumin inspired imidazo[1,2-a]pyridine analogues as tubulin polymerization Inhibitors. *Eur. J. Med. Chem.*, *143*, 216.

[39] Kumar, N. P., Sharma, P., Reddy,T. S., hankaraiah, N., Bhargava, S. K. &Kamal, A. (2018).Microwave-assisted one-pot synthesis of new phenanthrene fused-tetrahydrodibenzo-acridinones as potential cytotoxic and apoptosis inducing agents. *Eur. J. Med. Chem.*, *151*, 173.

[40] Ramya, P. V. S., Lalita, G., Srinivas, A., Shailaja, K., Digwal, C. S., Babu, B. N., Naidu, V. G. M. & Kamal, A. (2018). Curcumin inspired 2-chloro/phenoxy quinolone analogues: Synthesis and biological evaluation as potential anticancer agents. *Bioorg. Med. Chem. Lett.*, *28*, 892.

[41] Baig, M. F., Nayak, V. L., Prasad, B., Kishore, M., Satish, S., Gour, J. & Kamal, A. (2018). Synthesis and biological evaluation of imidazo[2,1-*b*]thiazole-benzimidazole conjugates as microtubule-targeting agents. *Bioorg. Chem.*, *77*,515.

[42] Kavita, D., Harikrishna, N., Shaik, S. P., Nanubolu, J. B., Shankaraiah, N. & Kamal, A. (2018). Iodine promoted dual oxidative C(sp3)-H amination of 2-methyl-3-arylquinazoline-4(3H)-ones; a facile route to 1, 4-dairylimidazo[1,5-*a*]quinazolin-5(4H)-ones. *Org. Biomol. Chem.*, *16*, 1720.

[43] Rahim, A., Shaik, S. P., Baig, M. F., Alarifi, A. & Kamal, A. (2018). Iodine mediated oxidative cross-coupling of unprotected anilines and heteroarylation of benzothiazoles with 2-methylquinoline. *Org. Biomol. Chem.*, *16*, 635.

[44] Sathish, M., Kavitha, B., Nayak, V. L., Tangella, Y., Ajitha, A., Nekkanti, S., Alarifi, A., Shankaraiah, N., Nagesh, N. & Kamal, A. (2018). Synthesis of podophyllotoxin linked β-carboline congeners as potential anticancer agents and DNA topoisomerase II inhibitors. *Eur. J. Med. Chem.*, *144*, 557.

[45] Reddy, V. G., Bonam, S. R., Reddy, T. S., Akunuri, R., Naidu, V. G. M., Nayak, V. L., Bhargava, S. K., Kumar, H. M. S., Srihari, P. & Kamal, A. (2018). 4β-Amidotriazole linked podophyllotoxin

congeners: DNA topoisomerase-IIα inhibition and potential anticancer agents for prostate cancer. *Eur. J. Med. Chem.*, *144*, 595.

[46] Mullagiri, K., Nayak, V. L., Satish, S., Mani, S. G., Guggilapu, S. D., Nagaraju, B., Alarifi, A. &Kamal, A. (2018). New (3-(1H-benzo[*d*]imidazol-2-yl)/(3-(3H-imidazo[4,5-*b*]pyridin-2-yl)-(1H-indol-5-yl)(3,4,5-trimethoxyphenyl)methanone conjugates as tubulin polymerization inhibitors. *Med. Chem. Comm.*, *9*, 275.

[47] Prashanth, S. A., Harikrishna, N., Hussaini, S. M. A., Babu, B. N. & Kamal, A. (2018).Glycogen synthase kinase-3 and its inhibitors: potential target for various therapeutic conditions. *Eur. J. Med. Chem.*, *144*, 843.

[48] Irfan, A., Rao, G. K., Makani, V. K. K., Shareef, M. A., Kumar, C. G., Bhadra, M. P. & Kamal, A. (2018). Design, synthesis and biological evaluation of 1,4-dihydroindeno[1,2-*c*]pyrazole linked oxindole analogues as potential anticancer agents targeting tubulin and inducing p53 dependent apoptosis. *Eur. J. Med. Chem.*, *144*, 104.

[49] Narsimha, R. M. P., Nagarju, B., Jeshma, K., Vishnuvardhan, M. V. P. S., Pollepalli, S., Jain, N. & Kamal, A. (2018). Synthesis of imidazo-thiadiazole linked indolinone conjugates and evaluated their microtubule network disrupting and apoptosis inducing ability. *Bioorg. Chem.*, *76*,420.

[50] Praveen, R. A., Sayeed, I. B., Shareef, M. A., Nagarju, B., Nagarajan, A. & Kamal, A. (2018). Development of pyrrolo [2,1-*c*][1,4]benzodiazepineβ-glucosideprodrugs for selective therapy of cancer. *Bioorg. Chem.*, *76*, 288.

[51] Jeshma, K., Nagarju, B., Nayak, V. L., Ravikumar, A., Rao, M. P. N., Nagesh, N. & Kamal, A. (2018). Design, synthesis and biological evaluation of new β-carboline-bisindole compounds as DNA binding, photocleavage agents and topoisomerase I inhibitors. *Eur. J. Med. Chem.*, *143*, 1563.

[52] Sultana, F., Bonam, S. R., Reddy, V. G., Nayak, V. L., Ravikumar, A., Routhu, S. R., Alarifi, A. & Kamal, A. (2018). Synthesis of benzo[*d*]imidazo[2,1-*b*]thiazole-chalcone conjugates as microtubule targeting and apoptosis inducing agents. *Bioorg. Chem.*, *76*, 1.

[53] Jeshma, K., Nagaraju, B., Kumar, C. J., Sirisha, K., Chandrasekhar, Ch., Alarifi, A., Kamal, A. (2018). Catalyst-free synthesis of pyrazole-aniline linked coumarin derivatives and their antimicrobial evaluation. *J. Saudi Chem. Soc.*, *22*, 665.

[54] Ramya, P. V. S., Angapelly, S., Digwal, C. S., Yadav, U., Babu, B. N. & Kamal, A. (2018). An efficient $RuCl_3 \cdot H_2O/I_2$ catalytic system: A facile access to 3-aroylimidazo[1,2-*a*]pyridines from 2-aminopyridines and chalcones. *J. Saudi Chem. Soc.*, *22*, 90.

[55] Ramya, P. V. S., Angapelly, S., Angeli, A., Digwal, C. S., Arifuddin, Md., Babu, B. N., Supuran, C. T. & Kamal, A. (2017). Discovery of curcumin inspired sulfonamide derivatives as a new class of carbonic anhydrase isoforms I, II, IX, and XII inhibitors. *J. Enzyme Inhbn. & Med. Chem.*, *32*, 1274.

[56] Seddigi, Z. S., Malik, M. S., Ahmed, S. A., Babalghith, A. O. & Kamal, A. (2017). Lipases in asymmetric transformations: Recent advances in classical kinetic resolution and lipase-metal combinations for dynamic processes. *Coord. Chem. Rev.*, *348*, 54.

[57] Dutta, S. G., Shaik, A. B., Kumar, C. G. & Kamal, A. (2017). Statistical optimization of production conditions of β-glucosidase from *Bacillus stratosphericus* strain SG9. *3 Biotech.*, *7*, 221.

[58] Ramya, P. V. S., Angapelly, S., Rani, R. S., Digwal, C. S., Kumar, C. G., Babu, B. N., Guntuku, L. & Kamal, A. (2017). Hypervalent iodine (III) catalyzed rapid and efficient access to benzimidazoles, benzothiazoles and quinoxalines: Biological evaluation of some new benzimidazole-imidazo[1,2-*a*]pyridine conjugates. *Arabian J. Chem.*, (*in press*).

[59] Angapelly, S., Ramya, P. V. S., Rani, R. S., Kumar, C. G., Kamal, A. & Arifuddin, Md. (2017). Ultrasound assisted, $VOSO_4$ catalyzed synthesis of 4-thiazolidinones: Antimicrobial evaluation of indazole-4-thiazolidinone derivatives. *Tetrahedron*, *58*, 4632.

[60] Shaik, A. B., Rao, G. K., Kumar, G. B., Patel, N., Reddy, V. S., Khan, I., Kumar, C. G., Veena, I., Shekar, K. C., Barkume, M., Jadhav, S., Juvekar, A., Kode, J., Bhadra, M. P. & Kamal, A. (2017).Design, synthesis and biological evaluation of novel pyrazolochalcones as

potential modulators of PI3K/Akt/mTOR pathway and inducers of apoptosis in breast cancer cells. *Eur. J. Med. Chem.*, *139*, 305.

[61] Nagarju, B., Jeshma, K., Nayak, V. L., Praveen, R. A., Alarifi, A. & Kamal, A. (2017). A facile one-pot CC and CN bond formation for the synthesis of spiro-benzodiazepines and their cytotoxicity. *Tetrahedron*, *73*, 6969.

[62] Kumar, N. P., Sharma, P., Kumari, S. S., Brahma, U., Nekkati, S., Shankariah, N. & Kamal, A. (2017). Synthesis of substituted phenanthrene-9-benzimidazole conjugates: Cytotoxicity evaluation and apoptosis inducing studies. *Eur. J. Med. Chem.*, *140*, 128.

[63] Satish, S., Shaik, P. S., Rao, A. V. S., Harikrishna, N., Alarif, A. & Kamal, A. (2017). Molecular iodine-promoted transimination for the synthesis of 6-phenylpyrido[2′,1′:2,3]imidazo[4,5-c]quinoline and 6-(pyridin-2-yl)pyrido[2′,1′:2,3]imidazo[4,5-c]quinolones. *Asian J. Org. Chem.*, *6*, 1830.

[64] Yadav, K., Meka, P. N. R., Guggilapu, S. S. S. D., Jeshma, K., Kamal, A., Srinivas, R., Devayani, P., Babu, B. N. & Nagesh, N. (2017). Telomerase inhibition and human telomeric G-quadruplex DNA stabilization by a β-Carboline-benzimidazole derivative at low concentrations. *ACS Biochemistry*, *56*, 4392.

[65] Sultana, F., Shaik, P. S., Nayak, V. L., Hussaini, S. M. A., Marumudi, K., Shaik, T. B., Sridevi, B., Alarifi, A. & Kamal, A. (2017). Design, synthesis and biological evaluation of 2-anilinopyridyl-linked oxindole conjugates as potent tubulin polymerisation inhibitors. *Chemistry Select*, *2*, 9901.

[66] Seddigi, Z. S., Malik, M. S., Prashanth, S. A., Ahmed, S., Babalghith, A. O., Lamfon, H. A. & Kamal, A. (2017). Recent advances in combretastatins based derivatives and prodrugs as antimitotic agents. *Med. Chem. Commun.*, *8*, 1592.

[67] Prasad, B., Reddy, V. G., Harikrishna, N., Subba Reddy, N. V., Alarifi, A. & Kamal, A. (2017). Annulation of 4-hydroxypyrones and α-keto vinyl azides; A regiospecific approach towards the synthesis of furo[3,2-c]pyrone scaffolds under catalyst free condition. *Chemistry Select*, *2*, 8122.

[68] Praveen R. A., Sastry, K. N. V., Jeshma, K., Reddy, V. G., Nagarajan, A., Sayeed, I. B. & Kamal, A. (2017).Visible light driven coupling of 2-aminopyridines and α-keto vinyl azides for the synthesis of imidazo[1,2-*a*]pyridines and their cytotoxicity. *Chemistry Select*, *2*, 8158.

[69] Tangella, Y., Manasa, K. L., Sathish, M., Nayak, V. L., Alarifi, A., Nagesh, A. & Kamal, A. (2017).An efficient one-pot approach for the regio and diastereoselective synthesis of trans-dihydrofuran derivatives: cytotoxicity and DNA-binding studies. *Org. Biomol. Chem.*, *15*, 6837.

[70] Sayeed, I. B., Rao, G. K., Makani, V. K. K., Nagarajan, A., Shareef, M. A., Alarifi, A., Bhadra, M. P. & Kamal, A. (2017). Development and biological evaluation of imidazothiazole propenones as tubulin inhibitors that effectively triggered apoptotic cell death in alveolar lung cancer cell line. *Chemistry Select*, *2*, 6480.

[71] Geeta, S. M., Shaik, P. S., Tangella, Y., Bale, S., Godugu, C. & Kamal, A. (2017). A facile I_2-catalyzed synthesis of imidazo[1,2-*a*]pyridines via sp^3 C-H functionalization of azaarenes and evaluation of anticancer activity. *Org. Biomol. Chem.*, *15*, 6780.

[72] Digwal, C. S., Yadav, U., Ramya, P. V. S., Sana, S., Swain, B. & Kamal, A. (2017). Vanadium-catalyzed oxidative C(CO)C(CO) bond cleavage for CN bond formation: One-pot domino transformation of 1,2-diketones and amidines into imides and amides. *J. Org. Chem.*, *82*, 7332.

[73] Vishnuvardhan, M. V. P. S., Saidireddy, V., Chandrashekar, V., Nayak, V. L., Sayeed, I. B., Alarifi, A. & Kamal, A. (2017). Click chemistry-assisted synthesis of triazolo linked podophyllotoxin conjugates as tubulin polymerization inhibitors. *Med. Chem. Commun.*, *8*, 1817.

[74] Bagul, C., Rao, G. K., Makani, K. K. V., Tamboli, J. R., Bhadra, M. P. & Kamal, A. (2017). Synthesis and biological evaluation of chalcone-linked pyrazolo[1,5-a]pyrimidines as potential anticancer agents. *Med. Chem. Commun.*, *8*, 1810.

[75] Sastry, K. N. V., Prasad, B., Nagaraju, B., Reddy, V. G., Alarifi, A., Babu, B. N. & Kamal, A. (2017). Copper-catalysed tandem synthesis of substituted quinazolines from phenacyl azides and o-carbonyl anilines. *Chemistry Select*, *2*, 5378.

[76] Rao, N. S., Shaik, A. B., Routhu, S. R., Hussaini, S. M. A., Sunkari, S., Rao, A. V. S., Reddy, A. M., Alarifi, A. & Kamal, A. (2017). New quinoline linked chalcone and pyrazoline conjugates: Molecular properties prediction, antimicrobial and antitubercular activities. *Chemistry Select*, *2*, 2989.

[77] Sultana, F., Shaik, S. P., Alarifi, A., Srivastava, A. K. & Kamal, A. (2017). Transition-metal-free oxidative cross-coupling of methylhetarenes with imidazoheterocycles towards efficient $C(sp^2)$–H carbonylation. *Asian J. Org. Chem.*, *6*, 890.

[78] Tangella, Y., Manasa, K. L., Sathish, M., Alarifi, A. & Kamal, A. (2017). Diphenylphosphoryl azide (DPPA)-mediated one-pot synthesis of oxazolo[4,5-c][1,8]naphthyridin-4(5H)-ones,oxazolo[4,5-c] quinolone-4(5H)-ones, and tosyloxazol-5-yl pyridines. *Asian J. Org. Chem.*, *6*, 898.

[79] Ramya, P. V. S., Angapelly, S., Babu, B. N., Digwal, C. S., Nagarsenkar, A., Srinivasulu, G., Prasanth, B., Arifuddin, Md., Kanneboina, K., Rangan, K. & Kamal, A. (2017). Metal-free C–C bond cleavage: One-pot access to 1,4-benzoquinone linked N-formyl amides/sulfonamides/carbamates using oxone. *Asian J. Org. Chem.*, *6*, 1008.

[80] Baig, M. F., Shaik, P. S., Krishna, H. N., Chouhan, K. N., Alarifi, A. & Kamal, A. (2017). One pot synthesis of naphtho [1',2':4,5] imidazo[1,2-a]pyridin-5-yl(aryl)methanones via a sequential Sonogashira coupling/alkyne-carbonyl metathesis. *Eur. J. Org. Chem.*, 4026.

[81] Mahesh, R., Nayak, V. L., Babu, K. S., Riyaz, S., Shaik, T. B., Kumar, G. B., Mallipeddi, P. L., Reddy, C. R., Shekar, K. C., Jose, J., Nagesh, N. & Kamal, A. (2017). Design, synthesis, and *in vitro* and *in-vivo* evaluations of (Z)-3,4,5-trimethoxystyryl benzenesulfonamides/

sulfonates as highly potent tubulin polymerization inhibitors. *Chem Med Chem.*, *12*, 678.

[82] Thoukhir, B. S., Hussaini, S. M. A., Nayak, V. L., Sucharitha, M. L., Malik, M. S. & Kamal, A. (2017). Rational design and synthesis of 2-an ilinopyridinyl benzothiazole schiff bases as antimitotic agents. *Bioorg. Med. Chem. Lett.*, *27*, 2549.

[83] Shaik, P. S., Vishnuvardhan, M. V. P. S., Sultana, F., Rao, A. V. S., Bagul, C., Bhattacharjee, D., Kapure, J. S., Jain, N. & Kamal, A. (2017). Design and synthesis of 1,2,3-triazolo linked benzo[d]imidazo[2,1-b]thiazole conjugates as tubulin polymerization inhibitors. *Bioorg. Med. Chem.*, *25*, 3285.

[84] Meshram, H. M., Sridevi, B., Tangella, Y., Korrapati, S. B., Nanubolu, J. B., Routhu, S. R., Kumar, C. G. & Kamal, A. (2017). Sulfamic acid catalyzed one-pot, three-component green approach: synthesis and cytotoxic evaluation of pyrazolyl-thiazole congeners. *New J. Chem.*, *41*, 3745.

[85] Reddy, N. C., Sathish, M., Adhikary, S., Nanubolu, J. B., Alarifi, A., Maurya, R. A. & Kamal, A. (2017). Phenacyl azides as efficient intermediates: one-pot synthesis of pyrrolidines and imidazoles. *Org. Biomol. Chem. 15*, 2730.

[86] Sayeed, I. B., Nayak, V. L., Shareef, M. A., Chouhan, N. K. & Kamal, A. (2017). Design, synthesis and biological evaluation of imidazopyridine-propenone conjugates as potent tubulin inhibitors. *Med. Chem. Comm.*, *8*, 1000.

[87] Rao, A. V. S., Rao, B. B., Sunkari, S., Shaik, P. S., Shaik, B. & Kamal, A. (2017). 2-Arylaminobenzothiazole-arylpropenone conjugates as tubulin polymerization inhibitors. *Med. Chem. Comm. 8*, 924.

[88] Ramya, P. V. S., Angapelly, S., Guntuku, L., Digwal, C. S., Babu, B. N., Naidu, V. G. M. & Kamal, A. (2017). Synthesis and biological evaluation of curcumin inspired indole analogues as tubulin polymerization inhibitors. *Eur. J. Med. Chem. 127*,100.

[89] Nayak, V. L., Nagesh, N., Ravikumar, A., Bagul, C., Vishnuvardhan, M. V. P. S., Srinivasulu, V. & Kamal, A. (2017). 2-aryl benzimidazole

conjugate induced apoptosis in human breast cancer MCF-7 cells through caspase independent pathway. *Apoptosis, 22*, 118.

[90] Rao, A. V. S., Swapna, K., Shaik, P. S., Nayak, V. L., Reddy, T. S., Sunkari, S., Shaik, T. S., Bagul, C. & Kamal, A. (2017).Synthesis and biological evaluation of *cis*-restricted triazole/tetrazole mimics of combretastatin-benzothiazole hybrids as tubulin polymerization inhibitors and apoptosis inducers. *Bioorg. Med. Chem., 25*, 977.

[91] Kumar, N. P., Sharma, T. P., Reddy, S., Nekkanti, S., Shankaraiah, N., Lalita, G., Sujanakumari, S., Bhargava, S. K., Naidu, V. G. M. & Kamal, A. (2017). Synthesis of 2,3,6,7-tetramethoxyphenanthren-9-amine: An efficient precursor to access new 4-aza-2,3-dihydropyrido phenanthrenes as apoptosis inducing agents. *Eur. J. Med. Chem., 127*, 305.

[92] Reddy, C. N., Krishna, N. H., Reddy, V. G., Alarifi, A. & Kamal, A. (2017). Metal-free aerobic oxidative C−C bond cleavage between the carbonyl carbon and the α-carbon of α-azido ketones: a novel synthesis of N-alkylated benzamides. *Asian J. Org. Chem., 6*, 1498.

[93] Baig, M. F., Shaik, S. P., Nayak, V. L., Alarifi, A. & Kamal, A. (2017). Iodine-catalyzed C_{sp3}-H functionalization of methylhetarenes: One-pot synthesis and cytotoxic evaluation of heteroarenyl-benzimidazoles and benzothiazole. *Bioorg. Med. Chem., 27*, 4039.

[94] Shaik, S. P., Sultana, F., Ravikumar, A., Sunkari, S., Alarifi, A. & Kamal, A. (2017). Regioselective oxidative cross-coupling of benzo[*d*]imidazo[2,1-*b*]thiazoles with styrenes: a novel route to C3-dicarbonylation. *Org. Biomol. Chem., 15*, 7696.

[95] Shaik, P. S., Nayak, V. L., Sultana, F., Rao, A. V. S., Shaik, A. B., Babu, K. S. & Kamal, A. (2017).Design and synthesis of imidazo[2,1-*b*]thiazole linked triazole conjugates: Microtubule-destabilizing agents. *Eur. J. Med. Chem., 126*,36.

Syed Shafi

Affiliation: Assistant Professor, Jamia Hamdard, New Delhi 110062, India since August 2015.

Education:

- Post-Doc. Research (Organometallic Chemistry), Institute of Organic Chemistry, Polish Academy of Sciences, Warsaw, Poland. 2009-2010.
- PhD (Chemistry), Sri Krishna Devaraya University-Anantapur, 2009(worked at IIIM-Jammu)
- MSc (Organic Chemistry), Sri Krishna Devaraya University-Anantapur, 2003
- BSc (Chemistry, Physics, Zoology), Sri Venkateswara University, Tirupati, 2001

Research and Professional Experience:

Present Research Interests: Multi-disciplinary research programmes including organic synthesis, organometallic chemistry, solid phase peptide synthesis and medicinal chemistry.

The thrust area of our research group is natural product inspired diversity oriented synthesis of bioactive ligands

Brief Report of previous Research:

- Synthesis of various new generation Grubb's catalysts, and their applications in bioactive molecules. We developed a novel tandem route for the generation of Pyrroles using Crossmetathesis.
- Reactions of resonance stabilised Grignard reagents/Indium halides to 1,3-dipolar agents: Domino synthesis of 1,5-disubstituted 1,2,3-triazoles and 3,5-disubstituted isoxazoles.
- Synthesis of bioactive heterocyclic compounds as small molecular high affinity ligands and the development of new methodologies.
- Design and synthesis of Non-steroidal anti-inflammatory agents (NSAIDs)
- Natural product inspired synthesis of antitubulin agents.

- Bis-heterocyclic conjugates of 1,2,3-triazoles/isoxazolines with bioactive natural products/biologically important ligands like artemisinin/2-mercapto benzothiazoles/indoleglyoxalic acid/ dibromopyrroles have been developed.
- Some new chemical reagents have been identified for the side chain cleavage of prednesilone.
- A new method has been developed for the deacylation of O-acylates.
- Development of novel immune adjuvants, in which a modified method was developed for the generation of TLR-2 specific Pam2Cys analogs in an economically favorable way. Some of analogues have shown promising adjuvant activity including Th1 and IL-12 responses, which is prerequisite for cell mediated immunity. In this part solid phase synthesis of Lysinyllysine based tripeptide analogs has also been carried out to explore the basic requirement and the minimal structural requirements for a peptide amphiphile to act as a better adjuvant candidate.
- Development of novel Domino reactions for Synthesis of bio-active molecules like 1,5-disubstituted 1,2,3-triazoles, 5-methyl isoxazoles, *bis*-heterocycles encompassing triazole and isoxazoles separated by two carbon spacer and Benzaldehyde-*O*-(2-cyclopropylidine-ethyl)-oximes (in water under green chemistry conditions).
- Apart from my thesis work, I have actively participated in the development of novel liquid phase (PEG-bound) methodologies (like Barbierallylation followed by Prins reaction and Ferrier rearrangement reaction) and Solution phase methodologies including carbohydrate based methodologies.
- As part of CSIR FYP supra institutional project at IIIM, I also have worked on solution phase methodologies and Picrosides and Betulin/Betulinic acid and Acteoside based scaffold modifications towards the development of novel target specific anticancer therapeutics.

Professional Appointments:

August 21, 2015 till date — Assistant Professor, Jamia Hamdard, New Delhi, India

2012 – 15 — DST-Fast Track Young Scientist at the Department of Chemistry, Jamia Hamdard, New Delhi-62. 2012-2015.

2010-12 — Hamdard Centenary Research Fellow (Postdoctoral) at the Department of Chemistry, Jamia Hamdard, New Delhi-62. 2010-2012.

2009-10 — Kasa Mianowskii postdoctoral fellow. Institute of Organic Chemistry, Polish Academy of Sciences, Warsaw, Poland.

2004-09 — CSIR -Junior/Senior Research Fellow

Honours

- Kasa-Mianowskii Postdoctoral Fellowship (Polish National Fellowship) awarded by Kasa-Mianowskii Foundation-Poland 2009.
- First Hamdard Centenary Research Fellowship from Jamia Hamdard 2010.
- Received best paper award from Jamia Hamdard for publication in high impact factor journal 2012.
- Received best paper award from Jamia Hamdard for publication in high impact factor journal 2014.

Publications from the Last 3 Years:

[1] Naaz, F., Ahmad, F., Lone, B. A., Pokharel, Y., Fuloria, N., Fuloria, S., Ravichandran, M., Lalithapattabhiraman, Shafi, Syed., & Yar, M. S. (2019). Design and synthesis of newer 1,3,4-oxadiazole and 1,2,4-triazole based Top sentin analogs as anti-proliferative agents targeting tubulin. *Bioorganic Chemistry*. (In press).

[2] Naaz, F., Haider, M. R., Syed, S., & Yar, M. S. (2019). Anti-tubulin agents of natural origin: Targeting taxol, vinca, and colchicine binding domains. *European Journal of Medicinal Chemistry*, 171, 310.

[3] Pratap, S., Naaz, F., Naaz., Reddy, S., Kunal, K. Jha., Sharma, K., Sahal, D., Akhter, Mymoona., Nayakanti, D., Kumar, M. S. H., Kumari, V., Pandey, K., & Shafi, Syed. (2018). Anti-proliferative and anti-malarial activities of spiroisoxazoline analogues of artemisinin. *Archiv Der Pharmazie, 1*.

[4] F., Naaz., Pallavi, M. C P., Shafi, S., Mulakayala, N., Yar, M. S. & Kumar, H. M. S. (2018). 1,2,3-triazole tethered Indole-3-glyoxamide derivatives as multiple inhibitors of 5-LOX, COX-2 & tubulin: Their anti-proliferative & anti-inflammatory activity. *Bioorganic Chemistry*, 81, 1.

[5] Siddiqui, M. A., Sattigeri, A. J., Shafi, S., Shamim, M., Singhal, S., & Malik, M. Z. (2018). Design, synthesis and biological evaluation of spiropyrimidinetrionesoxazolidinone derivatives as antibacterial agents. *Bioorganic & Medicinal Chemistry Letters*, 28, 1198.

[6] Karan, S., Kashyap, K. V., Shafi, S., & Saxena, K. A. (2017). Structural and inhibition analysis of novel sulfur-rich 2-mercaptobenzothiazole and 1,2,3-triazole ligands against Mycobacterium tuberculosis DprE1 enzyme. *Journal of Molecular Modelling,* 23, 241.

In: Benzothiazole
Editor: Atakan Heijstek

ISBN: 978-1-53617-548-6
© 2020 Nova Science Publishers, Inc.

Chapter 2

BENZATHIAZOLE ANALOGS AS ANTICONVULSANT AND ANTICANCER AGENTS

Ajayrajsinh R. Zala, Azazahemad Kureshi and Premlata Kumari*

Department of Applied Chemistry,
S. V. National Institute of Technology, Surat, Gujarat, India

ABSTRACT

Benzothiazole is an aromatic bicyclic heterocycle with 1, 3-thiazole fused with a benzene ring. On account of its extended π-delocalization, it binds to DNA molecules through π-π interaction and exhibits a wide assortment of dynamic destinations. Owing to the bioorganic and medicinal significance, the benzothiazole moiety would be possibly giving dynamic pharmacophores with diversified activity. Benzothiazole derivatives were found to show adequacy against some intense illnesses like cancer, neurodegeneration, neuropathic torment, infectious diseases, epilepsy, etc. As it contains electronegative atoms, it is achievable for different moieties to form conjugates. Riluzole, 2-Amino-6-trifluoro methoxy benzothiazole hydrochloride is an anticonvulsant drug that can also inhibit GABA (gamma-aminobutyric acid) uptake with an IC_{50} of 43 µM. Porters, 5-fluoro benzothiazole derivatives are an antitumor agent in phase I clinical trial that elicits selective activity against human-derived carcinomas of the breast, ovarian and renal origin. Considering the versatile nature of benzothiazole, various benzothiazole derivatives with anticonvulsant and anticancer activity will be discussed in this book chapter.

Keywords: benzothiazole, anticonvulsant, anticancer

1. INTRODUCTION

Heterocycles play a very significant role in medicinal chemistry. The majority of drug molecules exhibit therapeutic activity because of the heterocyclic scaffold. Little change in the heterocyclic moiety shows a major impact on the therapeutic activity of bioactive molecules. Benzothiazole is an aromatic bicyclic heterocycle consisting of two heteroatoms, Nitrogen and Sulfur, with 1, 3-thiazole fused with a benzene ring which makes the molecule active for a different functional group. As a heterocycle, Benzothiazole attracted medicinal chemists due to its diversified potential for biological efficacies [1]. On account of its extended π-delocalization, it binds to DNA molecules through π-π interaction and exhibits a wide assortment of dynamic destinations in the biological system. Owing to the bioorganic and medicinal significance, the Benzothiazole moiety would be possibly giving dynamic pharmacophores with diversified activity. Benzothiazole derivatives were found to show adequacy against some intense illnesses like cancer [2], neurodegeneration [3], neuropathic torment [4], infectious diseases [5], epilepsy [6], etc. Riluzole, 2-Amino-6-trifluoro methoxy benzothiazole hydrochloride is an anticonvulsant drug that can also inhibit GABA (gamma-aminobutyric acid) uptake with an IC_{50} of 43 µM. Porters, 5-fluoro benzothiazole derivatives are an antitumor agent. That means benzothiazole moieties containing compound shows numerous biological activities. Herewith we are introducing commonly formulated benzothiazole derivatives accounting various therapeutic properties.

Benzathiazole Analogs as Anticonvulsant and Anticancer Agents 73

1.1. Pharmacological Activity of Some Benzothiazole Derivatives

1.1.1. Antimicrobial Activity

Antimicrobial studies of (*E*)-5-(1-(benzo [*d*] thiazol-2-ylimino) ethyl)-4-(furan-2-yl)-6-methyl-3, 4-dihydro pyrimidine-2 (1*H*)-thione **(1)** were ascertained that the these compounds accounted the moderate antibacterial activity against both Gram-positive and Gram-negative bacterial culture [7] and (*E*)-5-amino-6-(benzo [*d*] thiazol-2-yl)-2- (2-benzylidene hydrazineyl)-7- (4-chloro phenyl) pyrido [2, 3-d] pyrimidin-4 (3*H*)-one **(2)** derivatives were evaluated for their antibacterial activity against *Taphylococcus aureus, Escherichia coli, Klebsiella pneumonia, Pseudomonas aeruginosa,* and *Streptococcus pyogenes* and for their antifungal activity against *Aspergillus flavus, Aspergillus fumigatus, Candida albicans, Penicillium marneffei,* and *Mucor*. Derivatives **(2)** with a nitro substituent, with an amine substituent and with fluorine all in para position displayed higher antibacterial activity relative to ciprofloxacin and antifungal activity as well with respect to standard drug clotrimazole [8].

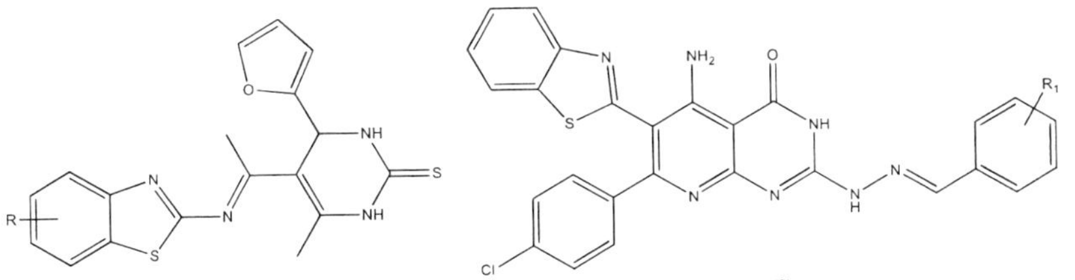

R= H, 6-OC$_2$H$_5$, 5-NO$_2$, 6-CH$_3$, 4-Cl   R$_1$= 4-NO$_2$, 4-NH$_2$, 4-F, 4-Cl, 2-Cl

Figure 1. Benzothiazole derivatives bearing with pyrimidine moiety [7, 8].

1.1.2. Anti-inflammatory Activity

Acetamide benzothiazole substituted derivatives are evaluated for anti-inflammatory activity. Their derivatives substituted with sulfonamide, Indole, 3-fluoro benzene, and furan showed significant anti-inflammatory activity. Among them, derivatives with a sulfonamide **(3)** and furan **(6)** substituents ascertained to be more active than the standard drug diclofenac sodium [9].

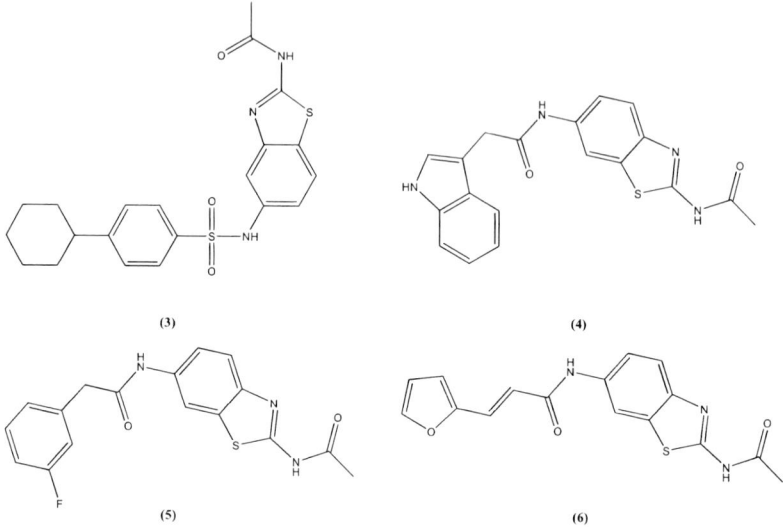

Figure 2. Benzothiazole derivatives bearing with acetamide moiety.

1.1.3. Antidiabetic Activity

AMP-activated protein kinase (AMPK) has recently emerged as a major potential target for novel antidiabetic drugs. Various 2-(((benzo [*d*] thiazol-2-yl) thio) methyl)-5-phenyl-1, 3, 4-oxadiazole **(7)** derivatives accounted for antidiabetic in alloxan-induced diabetic rat model [10]. All derivatives with a 6-nitro substituent on benzothiazole exhibited significant antidiabetic activity. Among them, derivatives with 5-*para* amino phenyl substituent on oxadiazole ring exerted maximum glucose-lowering effect. Likewise, 2-((benzo [*d*] thiazol-2-yl methyl) thio)-6-ethoxy benzo [*d*] thiazole **(8)** derivatives are screened for their antidiabetic activity also shown good activity [11].

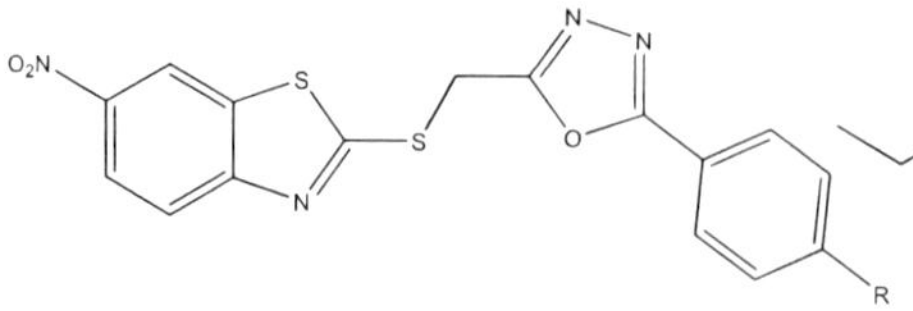

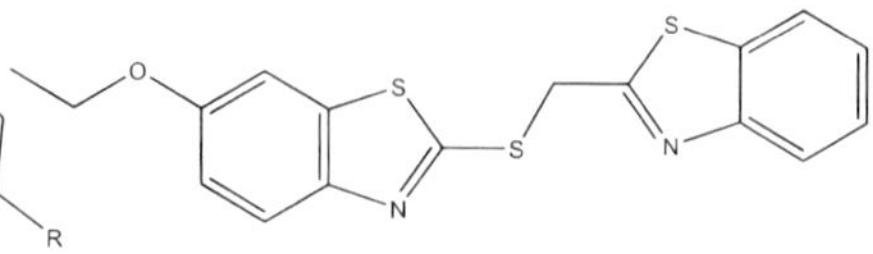

R = -C$_6$H$_5$, -p-NH$_2$C$_6$H$_4$-, p-NO$_2$C$_6$H$_4$-, p-OCH$_3$C$_6$H$_4$-

Figure 3. Benzothiazole derivatives bearing with oxadiazole moiety.

1.1.4. Antimalarial Activity

A series of benzothiazole hydrazones were accounted for its antimalarial activity *in-vitro* as well as *in-vivo*. Compound (*E*)-4-((2-(benzo [d] thiazol-2-yl) hydrazineylidene) methyl) benzene-1, 2-diol **(9)** was found to be the most active iron chelator as well as antiplasmodial. Against chloroquine/pyrimethamine resistant strain of *Plasmodium falciparum*. In *in-vivo* antimalarial activity also, compound **(9)** significantly suppressed the growth of multiple drug-resistant strain *Plasmodium yoelii*. Structure-activity relationship (SAR) studies ascertained that nitrogen and sulfur substituted five-membered aromatic ring present within the benzothiazole hydrazones might be responsible for their antimalarial activity [12].

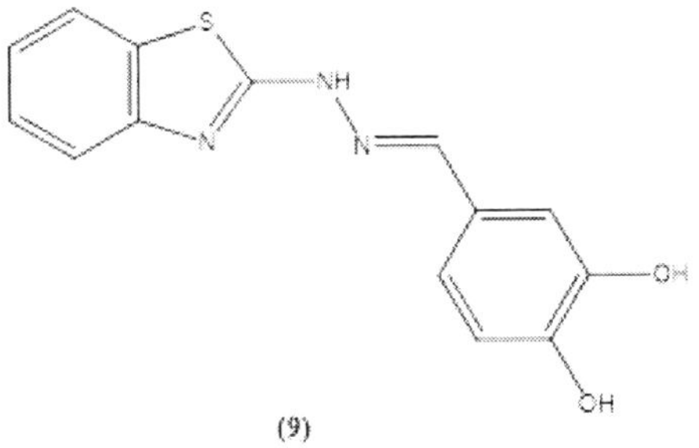

(9)

Figure 4. Structure of (E)-4-((2-(benzo [*d*] thiazol-2-yl) hydrazine ylidene) methyl) benzene-1, 2-diol.

1.1.5. Antioxidant Activity

Numerous 2-aryl benzothiazole derivatives were accounted for its antioxidant activity. The compound, **(10-13)** were found to be efficient scavengers of 2, 2-diphenyl-1-picryl hydrazyl (DPPH) and 2, 2'-azino-bis (3-ethyl benzothiazoline-6-sulphonic acid (ABTS) radicals. As per SAR, compound carrying $-OCH_3$, -COOH, -OH and $-NO_2$ group exerted appreciable antioxidant activity against both DPPH and ABTS radicals [13].

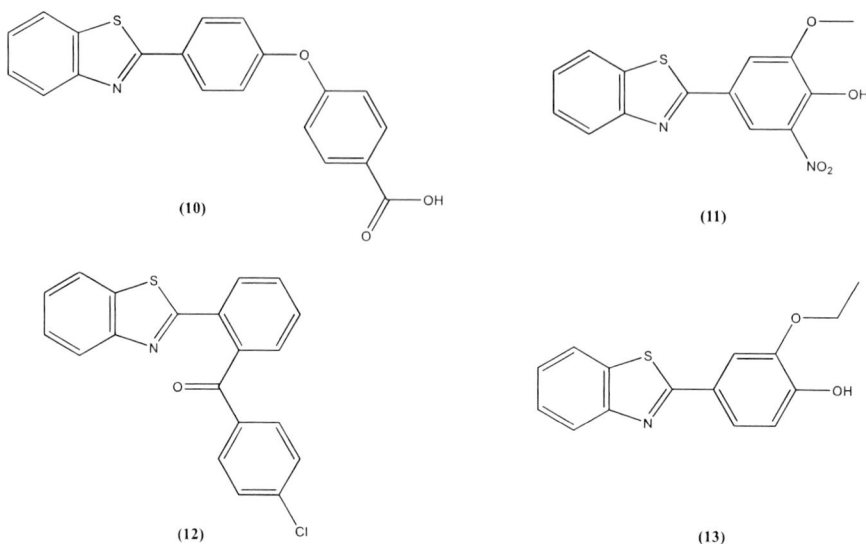

Figure 5. Structure of 2-aryl benzothiazole derivatives.

As benzothiazole contains electronegative atoms (N and S), it is achievable for different moieties to form conjugates. Benzothiazole containing molecules namely Benzothiazole semicarbazones derivatives [14] and 2-amino-6-trifluoro methoxy benzothiazole hydrochloride derivatives [15] shown to have anticonvulsant properties. Derivatives of 2-anilino benzothiazole [16], 2-substituted pyrimido Benzothiazole derivatives [17] and Benzothiazole acyl hydrazones derivatives [18] accounted to be antitumor agents. Considering the versatile nature of benzothiazole, we discussed in this book chapter anticonvulsant and anticancer activity of numerous benzothiazole derivatives.

2. ANTICONVULSANT ACTIVITY OF BENZOTHIAZOLE DERIVATIVES

Epilepsy is one of the most common neurological disease in accordance with WHO, which affects about 50 million people worldwide. It is estimated that up to 70% of people with epilepsy could live seizure-free if properly diagnosed and treated with appropriate antiseizure medicines. Despite of the availability of epileptic drugs, many patients with epilepsy do not encounter control in seizure and others encounter control on the cost of severe side effects [19]. Benzothiazole accounted to possess preliminary anticonvulsant properties [20]. So, numerous benzothiazole derivatives acquiring anticonvulsant activity are demonstrated. In recent years, benzothiazole derivatives have acquired conspicuous significance due to their wide spectrum of biological activities. Although they have been known for long to be biologically active [21, 22], their varied biological features are still of great scientific interest.

2.1. Benzothiazole Semicarbazones Derivatives

Clinically established anticonvulsant drugs demonstrated the presence of various hydrazones (=N-NH), amides (-CONH$_2$) and carbamides (-NHCO-NH-). In the same stream line, 1, 3-benzothiazole-2-yl semicarbazones accounted for their anticonvulsant, neurotoxicity and other toxic properties. The majority of the compounds were found to be active in MES (maximal electroshock test) screening. Four derivatives shown major activity against MES screening (E)-N-(6-chloro benzo [d] thiazol-2-yl)-2-(1-phenyl ethylidene) hydrazine-1-carboxamide **(14)**, (E)-N-(6-methyl benzo [d] thiazol-2-yl)-2-(1-(4-nitro phenyl) ethylidene) hydrazine-1-carboxamide **(15)**, (E)-N-(6-methoxy benzo [d] thiazol-2-yl)-2-(1-(4-nitro phenyl) ethylidene) hydrazine-1-carboxamide **(16)**, 2-(diphenyl methylene)-N-(6-methoxy benzo [d] thiazol-2-yl) hydrazine-1-carboxamide **(17)**. Their activity were comparable to drug phenytoin. All derivatives were screened at 30 mg/kg intraperitoneal injection in mice for anticonvulsant activity. Phenytoin was used as a standard drug at the dose 30 mg/kg. The derivatives **(14-17)** have given 100% protection at 0.5 h time intervals and compounds **(15-17)** have given 100% protection at 4 h time intervals except the compound **(14)** whose percentage protection decreased to 83.3% protection at 4 h time interval. The derivatives were evaluated for the neurotoxicity at the 30 mg/kg dose level; none of the derivatives shown the sign of neurotoxicity.

2.1.1. Synthesis of Benzothiazole Semicarbazone Derivatives (14-17)

The benzothiazole -2-yl urea **1a** were obtained by treating benzothiazole-2-amines with sodium cyanate in the presence of glacial acetic acid. The **1a** was then refluxed with hydrazine hydrate to form hydrazine carboxamides **1b**. Further **1b** was treated with appropriate ketones to get the final product **(14-17)** as per scheme 1 [14].

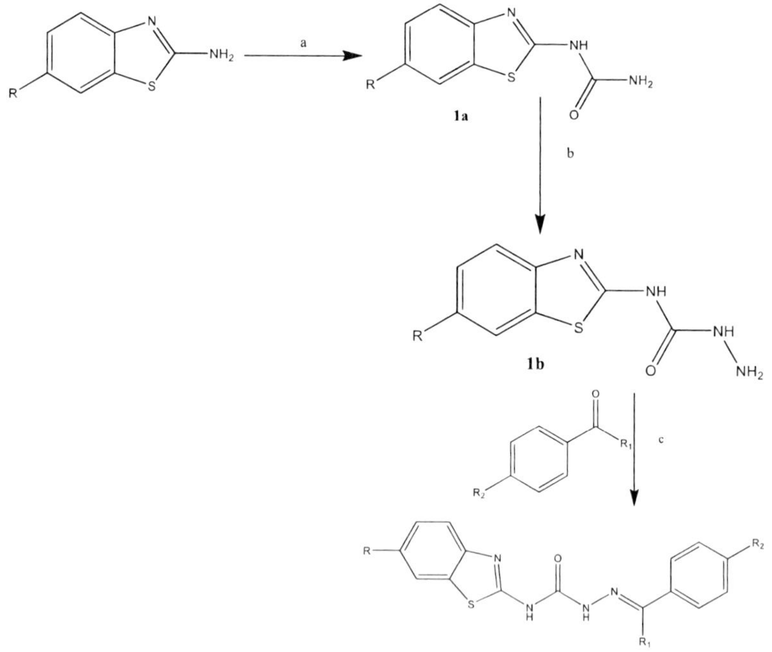

R= -Cl, -CH$_3$, -OCH$_3$; R$_1$= -CH$_3$, -C$_6$H$_5$; R$_2$= -NO$_2$, -OCH$_3$
Condition: (a) glacial acetic acid/ NaOCN; (b) NaOH/ EtOH, NH$_2$NH$_2$·H$_2$O, reflux 6 h; (c) EtOH/ reflux, 5 h.

Scheme 1. Synthesis of benzothiazole semicarbazones derivatives [14].

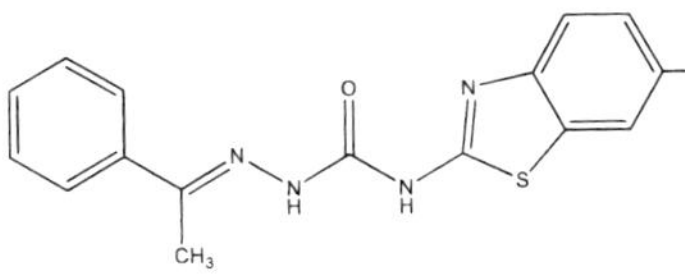

Figure 6. Structure of benzothiazole semicarbazones derivatives.

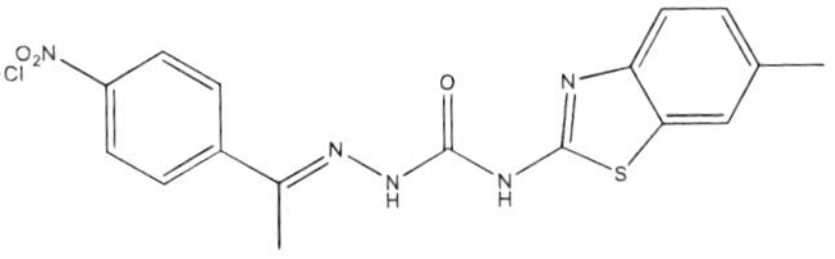

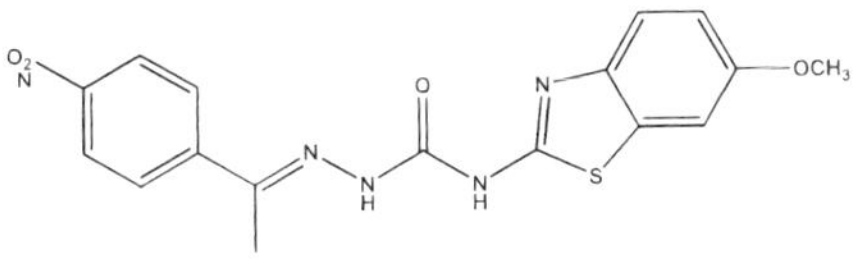

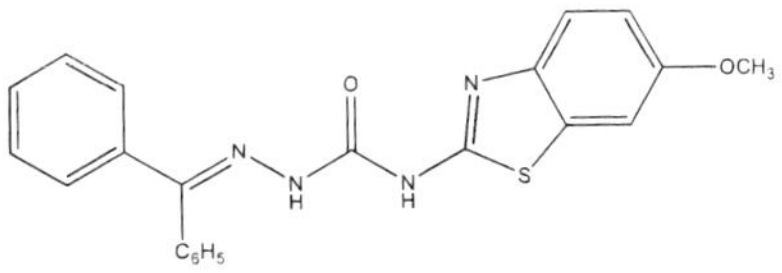

2.2. Benzothiazole Incorporated Barbituric Acid Derivatives

Phenobarbital and mephobarbital are a clinically available anticonvulsant drug in which barbituric acid is its essential part. In that streamline, two heterocycles, barbituric acid and benzothiazole possess anticonvulsant activity are hybridized together to obtain a series of 1-(6-substituted-1, 3-benzothiazole-2-yl)-3-(substituted phenyl) hexahydro-2, 4, 6-pyrimidinetriones. All the derivatives were evaluated for their anticonvulsant activity. Three of them 1-(6-chloro benzo [*d*] thiazol-2-yl)-3-(2-methoxy phenyl) pyrimidine-2, 4, 6 (1*H*, 3*H*, 5*H*)-trione **(18)**, 1-(6-chloro benzo [*d*] thiazol-2-yl)-3-(4-methoxy phenyl) pyrimidine-2, 4, 6(1*H*, 3*H*, 5*H*)-trione **(19)**, and 1-(4-methoxy phenyl)-3-(6-nitro benzo [*d*] thiazol-2-yl) pyrimidine-2, 4, 6(1*H*, 3*H*, 5*H*)-trio **(20)** showed promising anticonvulsant activities in MES (*maximal electroshock test*) screen test and subcutaneous pentylenetetrazole test (scPTZ). The standard drug used for this test Phenytoin, Ethosuximide and Phenobarbital. In the preliminary screening, 6 animals were used for the test and polyethylene glycol used as a solvent. 30, 100 and 300 mg/kg ratio doses form were administered.

The minimal amount of dose indicates whereby bioactivity was demonstrated in half or more of the mice. Examination time were 0.5 h and 4 h after injection was administered. All derivatives except few showed protection against MES test indicating these derivatives to forbid the seizure spread. At 30 mg/kg dose, only three compound 1-(6-chloro benzo [*d*] thiazol-2-yl)-3-(2-methoxy phenyl) pyrimidine-2, 4, 6 (1*H*, 3*H*, 5*H*)-trione **(18)**, 1-(6-chloro benzo [*d*] thiazol-2-yl)-3-(4-methoxy phenyl) pyrimidine-2, 4, 6(1*H*, 3*H*, 5*H*)-trione **(19)**, and 1-(4-methoxy phenyl)-3-(6-nitro benzo [*d*] thiazol-2-yl) pyrimidine-2, 4, 6(1*H*, 3*H*, 5*H*)-trio **(20)** found to be protective against seizure at 0.5 h of administration. Only **(19)** continued to protect from seizure at the same dosage. Compound **(18)** and **(20)** were also active at 4 h but at the higher dose of 100mg/kg. These compound did not exhibit any neurotoxicity even at highest dose of 100 mg/kg. Further, phase-II quantitative anticonvulsant activity in mice was performed using 10 animals and polyethylene glycol as a solvent to obtain anticonvulsant activity (ED_{50}) and neurotoxicity (TD_{50}). The protective index of compound **(18)** and **(20)** showed lower value i.e., less neurotoxic than some standard drugs like Phenobarbital and Valproate.

2.2.2. Synthesis of Benzothiazole Barbituric Acid Derivatives (18-19)

Substituted anilines are treated with potassium thiocyanate and bromine in acetic acid to yield substituted **1a** derivatives. On refluxing it with ethyl chloroformate and triethylamine, these derivatives yielded the carbamate derivatives of Benzothiazole **1b**. Derivatives **1b**, further condensed with substituted aniline to get **1c**. Finally, **1c** were condensed with Malonic acid in the presence of acetyl chloride to afford the desired compounds **(18-20)** as per scheme 2 [15].

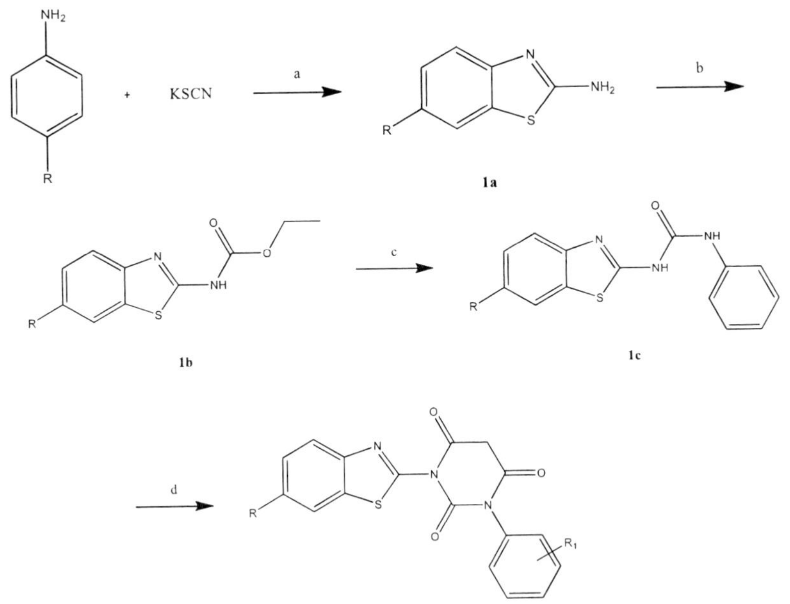

R = -Cl, -Br, -NO$_2$, -F R$_1$= -H, 2-CH$_3$, 2-OCH$_3$, 4-OCH$_3$
Condition: (a) Br$_2$, Acetic acid; (b) ClCO$_2$Et, triethyl amine, reflux; (c) R$_1$PhNH$_2$, reflux; (d) CH$_2$(CO$_2$H)$_2$, CH$_3$COCl, 40°C, 4 h.

Scheme 2. Synthesis of benzothiazole barbituric acid derivatives (18-20) [15].

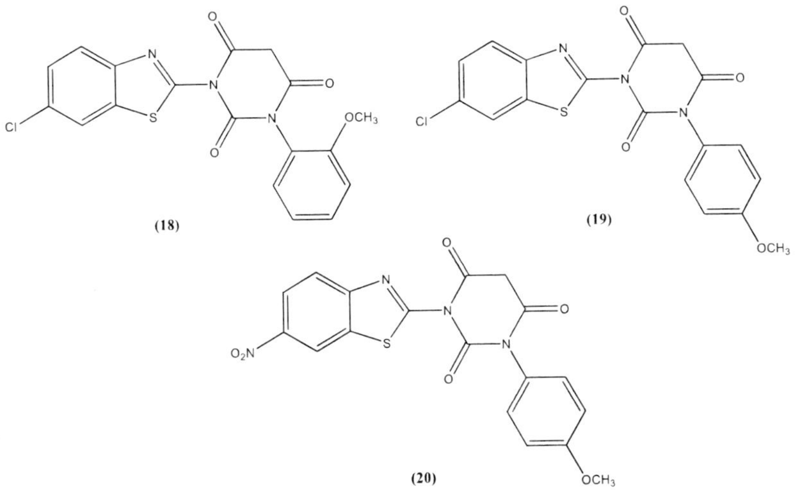

Figure 7. Structure of 2-aryl benzothiazole derivatives (18-20).

3. ANTICANCER ACTIVITY OF BENZOTHIAZOLE DERIVATIVES

In accordance with the GLOBOCAN 2018 database, the global cancer burden has risen to 18.1 million new cases and 9.6 million deaths in 2018. Nearly half of the new cases and more than half of the cancer deaths worldwide in 2018 are estimated to occur in Asia which acquires 60% of the global population. Even though Europe has 9% of the global population, it accounts for 23.4% global cancer cases and 20.3% of cancer deaths. Estimated data indicate that cancer is still one of the wildest spreading diseases and common reasons for deaths worldwide. In spite of the widespread research and fast changes in cancer therapy, there is still a need for new efficient treatment. The discovery of new chemotherapeutics is of the principal importance due to the essential capability of tumor cells to develop resistance to available agents. Numerous benzothiazoles are accounted for its potent anticancer activity. Some are discussed in this section.

3.1. 2-Anilino Benzothiazole Derivatives

Protein tyrosine kinases play a significant role in controlling cellular proliferation. Over expression of certain receptor tyrosine kinases play a crucial role in cancer development. Like the epidermal growth factor (EGR) receptor tyrosine kinases are frequently expressed in high levels in certain carcinomas particularly breast, colon, and bladder cancer [24]. Benzothiazoles act via striving against ATP for binding at the catalytic domain of tyrosine kinase [25]. Various 2-anilino benzothiazoles accounted as EGFR inhibitors. Both electron withdrawing (-Cl) and substituted at position 2 of electron donating (-CH$_3$) substituted at position 2 of anilines were incorporated to check their effect on anticancer activity. The tumor growth properties of the two compounds N-(2,6-dimethyl phenyl)-6-methyl benzo [*d*] thiazol-2-amine (**21**) and N-(2, 6-dichloro phenyl)-6-methyl benzo [*d*] thiazol-2-amine (**29**) were screened on human tumor cell lines at a single high dose (10^{-5} M) and five dose level by National cancer institute, NCI, USA.

Compound (**21**) shown remarkable anticancer activity against nine different subpanels with GI_{50} value against Leukemia SR cell line 4.12×10^{-7} M, Non-Small Cell Lung Cancer NCI-H5222 cell line against 5.52×10^{-7} M, Colon Cancer HCT-15 against 9.20×10^{-7} M, CNS cancer SF-295 cancer cell line against 7.25×10^{-7} M, Melanoma MDA-MB-435 cancer cell line against 2.42×10^{-7} M, Ovarian NCI/ADR-RES cancer cell line against 6.39×10^{-7} M, Renal Cancer CAKI-1 cancer cell line against 5.64×10^{-7} M and Breast MDA-MB-468 cancer cell line against 3.30×10^{-7} M.

Compound (**29**) shown incredible anticancer activity against nine different subpanels with GI_{50} value against Leukemia SR cell line 8.29×10^{-7} M, Non-Small Cell Lung Cancer NCI-H5222 cell line against 7.18×10^{-8} M, CNS cancer SF-295 cancer cell line against 4.59×10^{-7} M, Melanoma MDA-MB-435 cancer cell line against 8.13×10^{-7} M, Ovarian NCI/ADR-RES cancer cell line against 5.31×10^{-7} M, Renal Cancer CAKI-1 cancer cell line against 1.61×10^{-7} M, Prostate PC-3 cancer cell line against 5.26×10^{-7} and Breast MDA-MB-231/ATCC cancer cell line against 3.04×10^{-7} M. It shown the sensitivity against individual cell line and shown highest activity against non-small cell HOP-92 lung cancer cell line with GI_{50} value of 7.18×10^{-8} M. Compound (**29**) accounted to be a most active compound of this series. This benzothiazole analog could be considered as useful templates for future development to obtain more potent antitumor agents.

3.1.1. Synthesis of 2-Anilino Benzothiazole Derivatives (21-30)

Substituted thioureas [26] synthesized utilizing various substituted isothiocyanates [27] which were treated with bromine in chloroform to yield various 2-anilino benzothiazole derivatives (**21-30**) as per scheme 3 [16].

Table 1. R_1 and R_2 substituents in derivatives 2-Anilino Benzothiazole Derivatives (21-30)

Substituent	21	22	23	24	25	26	27	28	29	30
R_1	CH_3	Br	NO_2	H	H	CH_3	Br	NO_2	H	H
R_2	H	H	H	Cl	NO_2	H	H	H	Cl	NO_2

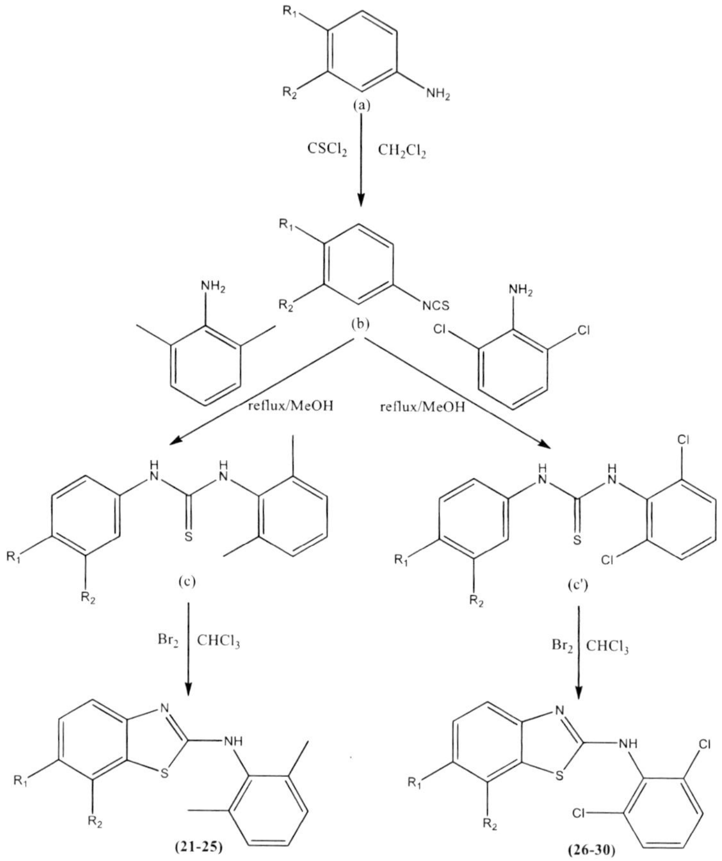

Scheme 3. Synthesis of 2-anilino benzothiazole derivatives (21-30) [16].

3.2. 2-Substituted Pyrimido Benzothiazole Derivatives

The pharmacological activity of 3-substituted 2-imino benzothiazole was found to be three-times more potent than Riluzole [28], a blocker of excitatory amino acids mediated neurotransmission [29]. Taking into consideration the importance of biological activities and various applications of pyrimidines, oxo pyrimidines, imino pyrimidines, amino, and imino benzothiazoles, numerous pyrimido benzothiazoles and its 2-substituted derivatives were accounted for their anticancer activity.

Derivatives **(31-36)** were selected by NCI through the Developmental Therapeutics Program (DTP) process and were screened for their *in vitro* anticancer activity towards 60 cancer cell lines at a single dose of 10 µM. DTP operates a repository of synthetic and pure natural products, which are evaluated as potential anticancer agents. All the Six compounds shown anticancer activity against 60 cell lines panels.

Compounds **(32)** and **(33)** expressed remarkable inhibition capacity againstHOP-92 (Lung cancer), UACC-62 (Melanoma), due to the presence of *p*-CH_3 and *p*-OCH_3 group. Compound **(34)** displayed excellent activity against k-562, RPMI-8226 (Leukemia cancer), HOP-92, UO-31 (Renal cancer) cell lines due to presence of *p*-Cl group. Compounds **(35)** and **(36)** presented inhibitory effect against CAKI-1, UO-31 (Renal cancer), MCF-7 (Breast cancer) and K-562 (Leukemia cancer), CAKI-1, UO-31 (Renal cancer), PC-3 (Prostate cancer) respectively due to the presence of heteryl cyclic amines at position-2 respectively.

3.2.1. Synthesis of Pyrimido benzothiazole and its 2-substituted Derivatives (32-36)

2-amino-4, 7-dimethyl benzothiazole, and bis- (methylthio) methylene malononitrile were refluxed together with anhydrous K_2CO_3 and dimethylformamide, DMF as a solvent to afford the compound 4-imino-6, 9-dimethyl-2-(methylthio)-4*H*-benzo [4, 5] thiazolo [3, 2-a] pyrimidine-3-carbonitrile **(31)**. Compound **31** was further refluxed with aromatic amines/ phenols/ heteroaryl amines in the presence of DMF and anhydrous K_2CO_3 to yield various pyrimido benzothiazole derivatives **(32-36)** as per Scheme 4 [17].

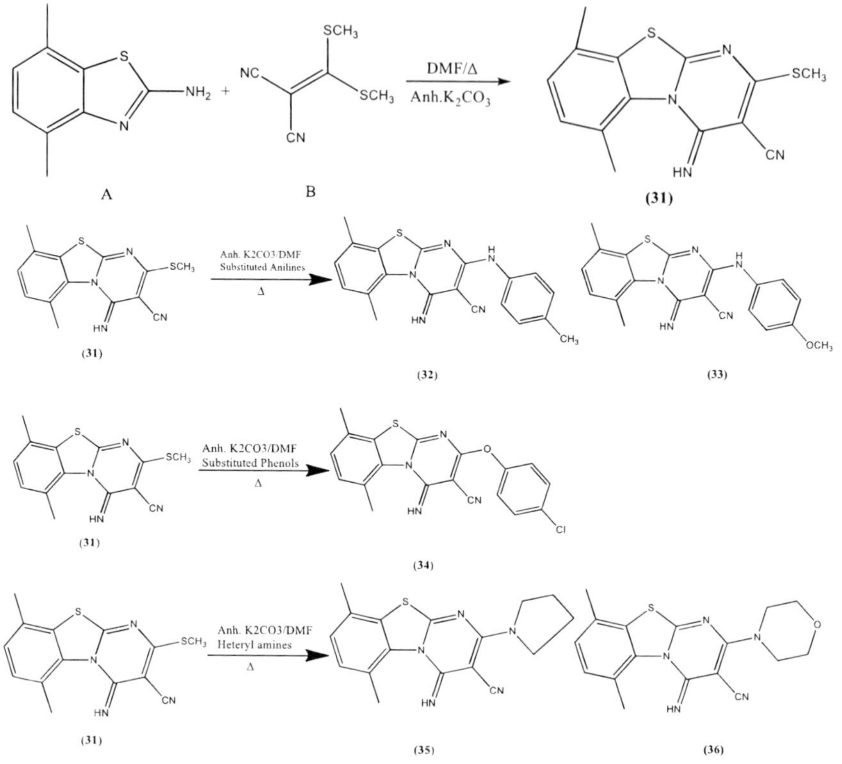

Scheme 4. Synthesis of 2-substituted pyrimido Benzothiazole derivatives [17].

3.3. Benzothiazole Acyl Hydrazones Derivatives

Various substitutions at specific positions on the Benzothiazole scaffold are known to regulate the antitumor property. N-acyl aryl hydrazone scaffold has been widely us notified as a building block in several antitumor agents due to its flexible skeleton and the presence of both hydrogen donors and acceptors [30]. Antitumor acyl hydrazones have been experienced to induce apoptosis in diverse carcinogenic cells. In addition, the compounds include benzothiazole and hydrazone moieties on the same structure that have significant anticancer activity [31, 32]. So numerous derivatives with acyl hydrazone moiety as a linker between benzothiazole and phenyl scaffold accounted for its anticancer activity. Antitumor potential of benzothiazole acyl hydrazones derivatives were assessed against A549 (human lung adenocarcinoma epithelial cell line), C_6 (rat brain glioma cell line), MCF-7(human breast adenocarcinoma cell line), and HT-29(human colorectal adenocarcinoma cell line) [33]. Derivatives (**37-46**) were also evaluated against healthy NIH3T3 (Mouse embryo fibroblast cell line) to study selectivity of these derivatives for carcinogenic cells. Compound (*E*)-2-((5-chloro benzo [*d*] thiazol-2-yl) thio)-N'-(4-(4-(4-methoxy phenyl) piperazin-1-yl) benzylidene) aceto hydrazide (**41**) was accounted for the highest cytotoxic activity against A549 cell lines. Furthermore, the IC_{50} value (0.03 mM) against A549 cell was half of that of cisplatin (0.06 mM).In contrast, another compound could not show significant activity against A549 cells. Compound (*E*) - 2- ((5-chloro benzo [*d*] thiazol-2-yl) thio)-N'-(4-(4-methyl piperidin-1-yl) benzylidene) acetohydrazide (**40**), (*E*)-2-((5-chloro benzo [*d*] thiazol-2-yl) thio)-N'-(4-(4-(4-methoxy phenyl) piperazin-1-yl) benzylidene) aceto hydrazide (**41**) and (*E*) - 2- ((5-methoxy benzo [*d*] thiazol-2-yl) thio)-N'-(4-(4-methyl piperidin-1-yl) benzylidene) acetohydrazide (**45**) were shown incredible cytotoxicity with IC_{50}, 0.03 mM same as cisplatin against C6 cell line. Compound (E)-2-((5-chloro benzo [d] thiazol-2-yl) thio)-N'-(4-(4-(4-methoxy phenyl)-3-methyl piperazin-1-yl) benzylidene) acetohydrazide (**39**) and (**40**) were notified to be most active against MCF-7 and HT-29 cell line. C_6 cell line was accounted to be more susceptible to these compounds in comparison with A 549, MCF-7 and HT-29 cell lines. DNA synthesis inhibition assay of C_6 cell lines shown (**40**) has higher inhibition potency than cisplatin at all concentration i.e., $IC_{50}/4$, $IC_{50}/2$, IC_{50}, $2 \times IC_{50}$ and $4 \times IC_{50}$. Compound (**41**) also displayed a higher antiproliferative effect than cisplatin but at concentration, $IC_{50}/2$, IC_{50}, $2 \times IC_{50}$ and $4 \times IC_{50}$. Flow cytometric analysis used to study the pathway of cell death indicating Compounds (**40**), (**41**), and (**44**) induced apoptotic cell death dose-dependently in the cancer cells. Out of these, compound (**41**) were shown most apoptotic activity (19.5% apoptotic cells) in comparison with cisplatin (20.9% apoptotic cells). These compounds did not kill healthy cells at concentration with cytotoxic effects against cancer cells depicting their selective inhibition towards cancer cells. Structurally, the 5-chloro substitution of benzothiazole increases the cytotoxicity against carcinogenic cell lines. Furthermore, comparing 2-, 3-, or 4- methyl substitution of piperidine depicted that 3- or 4- substituents contributing more to cytotoxicity against carcinogenic cell lines.

3.3.1. Synthesis of Benzothiazole Acyl Hydrazones Derivatives (37-46)

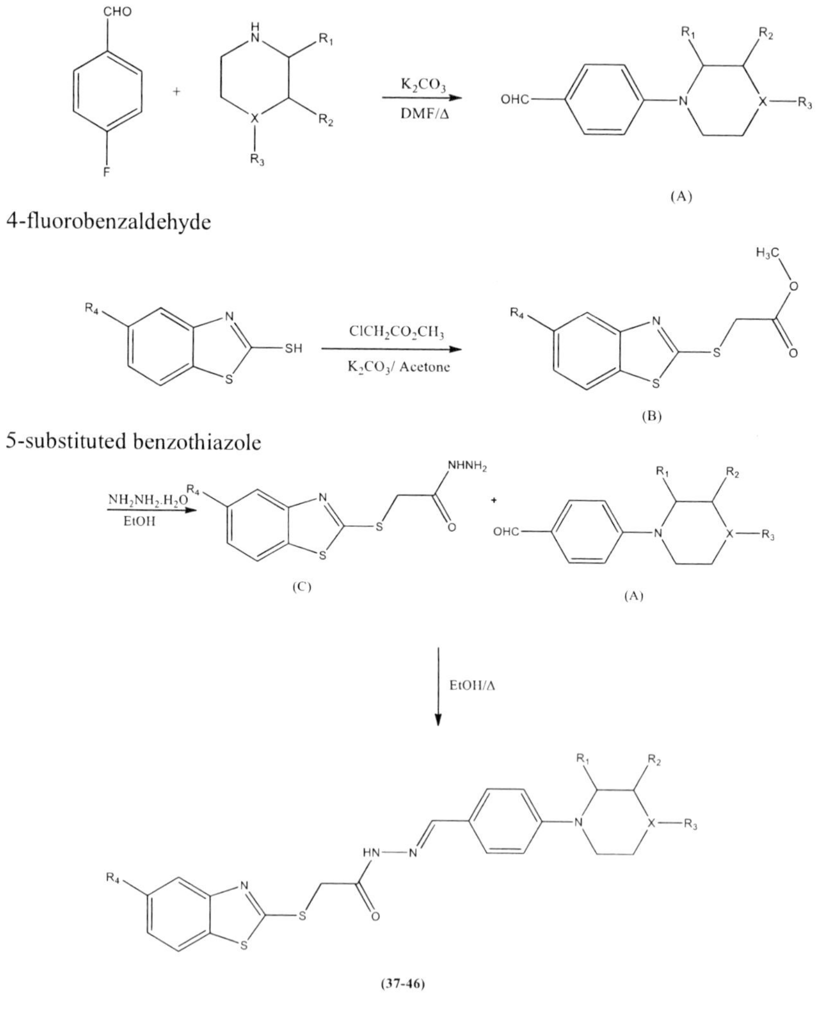

Scheme 5. Synthesis of benzothiazole acyl hydrazones derivatives (37-46) [18].

Benzothiazole acyl hydrazones derivatives **(37-46)** were synthesized through the Scheme 5. In the beginning, 4-substituted benzaldehydes **(A)** were synthesized via 4-fluoro benzaldehyde and appropriate secondary amine [34]. Subsequently, 5-substituted benzothiazole and ethyl chloro acetate yielded **(B)** which with an excess of hydrazine hydrate gave **(C)** [35]. Finally, **(C)** and **(A)** were reacted to afford benzothiazole acyl hydrazones derivatives **(37-46)** as per Scheme 5 [18].

Table 2. R_1, R_2, R_3, R_4 and X substituents in derivatives 2-Anilino Benzothiazole Derivatives (37-46)

Compound	R_1	R_2	R_3	R_4	X
(37)	-H	-H	-H	-Cl	-CH
(38)	-CH$_3$	-H	-H	-Cl	-CH
(39)	-H	-CH$_3$	-H	-Cl	-CH
(40)	-H	-H	-CH$_3$	-Cl	-CH
(41)	-H	-H	4-methoxy phenyl	-Cl	-N
(42)	-H	-H	-H	-OCH$_3$	-CH
(43)	-CH$_3$	-H	-H	-OCH$_3$	-CH
(44)	-H	-CH$_3$	-H	-OCH$_3$	-CH
(45)	-H	-H	-CH$_3$	-OCH$_3$	-CH
(46)	-H	-H	4-methoxy phenyl	-OCH$_3$	-N

CONCLUSION

Benzothiazole scaffold apprehended to be a dynamic pharmacophore with vast and versatile bioorganic and medical significance. This chapter highlighted synthetic routes of benzothiazole derivatives which accounted to have anticonvulsant and anticancer property. The discussed numerous benzothiazole derivatives could be further optimized to develop a lead compound of biological significance. Benzothiazole semicarbazones; (*E*)–N - (6-methyl benzo[*d*] thiazol-2-yl) -2-(1- (4-nitro phenyl) ethylidene) hydrazine-1-carboxamide **(15)**, (*E*)-N-(6-methoxy benzo[*d*] thiazol-2-yl) -2- (1- (4-nitro phenyl) ethylidene) hydrazine-1-carboxamide **(16)** and 2-

(diphenyl methylene)-N-(6-methoxy benzo[*d*] thiazol-2-yl) hydrazine-1-carboxamide **(17)** derivative shown 100% protection in MES same as the standard employed Phenytoin. In benzothiazole barbituric acid derivative, two active pharmacophores showing promising anticonvulsant activity 1-(6-chloro benzo [*d*] thiazol-2-yl)-3-(2-methoxy phenyl) pyrimidine-2, 4, 6 (1*H*, 3*H*, 5*H*)-trione **(18)** and 1- (6-chloro benzo [d] thiazol-2-yl)-3- (4-methoxy phenyl) pyrimidine-2, 4, 6 (1*H*, 3*H*, 5*H*)-trione **(19)**. The most active compounds were quantitatively observed for anticonvulsant activity (ED_{50}) and neurotoxicity (TD_{50}) and accounted to be comparable to standard drugs. 1- (6-chloro benzo [*d*] thiazol-2-yl)-3-(2-methoxy phenyl) pyrimidine-2, 4, 6 (1*H*, 3*H*, 5*H*)-trione **(18)** and 1-(4-methoxy phenyl)-3-(6-nitro benzo [*d*] thiazol-2-yl) pyrimidine-2, 4, 6 (1*H*, 3*H*, 5*H*)-trione **(20)** shows higher Protective Index than the standard drug-like Phenobarbital and Valproate. Among 2-anilino benzothiazole showing anticancer activity, the compound N-(2, 6-dichloro phenyl)-6-methyl benzo [*d*] thiazol-2-amine **(29)** against non-small cell HOP-92 lung cancer cell line proved to be the most active derivative. Derivatives of 3-substituted pyrimido benzothiazole showing promising different activity against Leukemia lung, Melanoma, CNS, Colon, Ovarian, Renal, Prostate, and Breast cancer cell lines. Benzothiazole acyl hydrazones indicated that (*E*) - 2- ((5-chloro benzo [*d*] thiazol-2-yl) thio)-N'-(4-(4-methyl piperidin-1-yl) benzylidene) acetohydrazide **(40)**, which is the lead compound of the series, possesses significant selective cytotoxic effects on cancer cells. Furthermore, the same compound **(40)** shows higher antiproliferative activity than cisplatin. The discussed derivatives would be significant for further research in the evolution of a better broad range of more potent anticonvulsant and anticancer agents.

REFERENCES

[1] Srivastava, A., Mishra, A. P., Chandra, S., and Bajpai, A. (2019). Benzothiazole derivative: a review on its pharmacological importance

towards synthesis of lead. *International Journal of Pharmaceutical Sciences and Research, 2 (10)*, 1553-1566.

[2] Vasconcelos, Z. S., Ralph, A. C. L., Calcagno, D. Q., dos Santos Barbosa, G., do Nascimento Pedrosa, T., Antony, L. P., and de Vasconcellos, M. C. (2018). Anticancer potential of benzothiazolic derivative (E)-2-((2-(benzo [d] thiazol-2-yl) hydrazono) methyl)-4-nitrophenol against melanoma cells. *Toxicology in Vitro, 50*, 225-235.

[3] Araki, T., Muramatsu, Y., Tanaka, K., Matsubara, M., and Imai, Y. (2001). Riluzole (2-amino-6-trifluoromethoxy benzothiazole) attenuates MPTP (1-methyl-4-phenyl-1, 2, 3, 6-tetrahydropyridine) neurotoxicity in mice. *Neuroscience letters, 312* (1), 50-54.

[4] Ramnauth, J., Rakhit, S., Maddaford, S., and Bhardwaj, N. (2006). *U.S. Patent No. 7,141,595*. Washington, DC: U.S. Patent and Trademark Office.

[5] Bailey, T. R., and Pevear, D. C. (2004). Benzothiazole compounds, compositions and methods for treatment and prophylaxis of rotavirus infections and associated diseases. *WO2004078115 A2, September*, 16.

[6] Romettino, S., Lazdunski, M., and Gottesmann, C. (1991). Anticonvulsant and sleep-waking influences of riluzole in a rat model of absence epilepsy. *European journal of pharmacology, 199 (3)*, 371-373.

[7] Shinde, P. K., and Waghamode, K. T. (2017). Synthesis, characterization and antibacterial activity of substituted benzothiazole derivatives. *International Journal of Scientific and Research Publications, 7 (8)*, 365-370.

[8] Maddila, S., Gorle, S., Seshadri, N., Lavanya, P., and Jonnalagadda, S. B. (2016). Synthesis, antibacterial and antifungal activity of novel benzothiazole pyrimidine derivatives. *Arabian Journal of Chemistry, 9* (5), 681-687.

[9] Sadhasivam, G., and Kulanthai, K. (2015). Synthesis, characterization, and evaluation of anti-inflammatory and anti-diabetic activity of new benzothiazole derivatives. *Journal of Chemical and Pharmaceutical Research, 7* (8), 425-431.

[10] Kumar, S., Rathore, D. S., Garg, G., Khatri, K., Saxena, R., and Sahu, S. K. (2016). Synthesis and evaluation of some 2-((benzo thiazol-2-ylthio) methyl)-5-phenyl-1, 3, 4-oxadiazole derivatives as antidiabetic agents. *Asian Pacific Journal of Health Sciences, 3* (4), 65-74.

[11] Meltzer-Mats, E., Babai-Shani, G., Pasternak, L., Uritsky, N., Getter, T., Viskind, O., and Gruzman, A. (2013). Synthesis and mechanism of hypoglycemic activity of benzothiazole derivatives. *Journal of medicinal chemistry, 56* (13), 5335-5350.

[12] Sarkar, S., Asim A., Siddiqui, A. A., Saha, S. J., De R., Mazumder, S., Banerjee, C., Iqbal M. S., Nag S., Adhikari S., and Bandyopadhyaya U. (2016). Antimalarial activity of small-molecule benzothiazole hydrazones. *Antimicrobial Agents and Chemotherapy, 60 (7)*, 4217-4228.

[13] Amin S., and Parle A. (2018). Synthesis, characterization and antioxidant activity of 2-aryl benzothiazole derivatives. *International Journal of Current Pharmaceutical Research, 10 (5)*, 3-8.

[14] Siddiqui, N., Rana, A., Khan, S. A., Bhat, M. A., and Haque, S. E. (2007). Synthesis of benzothiazole semicarbazones as novel anticonvulsants—The role of hydrophobic domain. *Bioorganic & medicinal chemistry letters, 17 (15)*, 4178-4182.

[15] Siddiqui, N., and Ahsan, W. (2009). Benzothiazole incorporated barbituric acid derivatives: synthesis and anticonvulsant screening. *Archiv der Pharmazie: An International Journal Pharmaceutical and Medicinal Chemistry, 342 (8)*, 462-468.

[16] Noolvi, M. N., Patel, H. M., and Kaur, M. (2012). Benzothiazoles: search for anticancer agents. *European journal of medicinal chemistry, 54*, 447-462.

[17] Waghmare, G. S., Chidrawar, A. B., Bhosale, V. N., Shendarkar, G. R., and Kuberkar, S. V. (2013). Synthesis and in-vitro anticancer activity of 3-cyano-6, 9-dimethyl-4-imino 2-methylthio 4H-pyrimido [2, 1-b] [1, 3] benzothiazole and its 2-substituted derivatives. *Journal of pharmacy research, 7* (9), 823-827.

[18] Osmaniye, D., Levent, S., Karaduman, A., Ilgın, S., Özkay, Y., and Kaplancıklı, Z. (2018). Synthesis of new benzothiazole acyl hydrazones as anticancer agents. *Molecules, 23* (5), 1054.

[19] Coatsworth, J. J. (1971). *Studies on the clinical efficacy of marketed antiepileptic drugs* (No. 12). National Institutes of Health; [for sale by the Supt. of Docs., US Govt. Print. Off., Washington].

[20] Siddiqui, M. (1992). Molecular Cloning and Regulation of Angiotensinogen Gene Expression. In *Recent Advances in Cellular and Molecular Biology: From the 1. World Congress of CMB, Paris, September 1-7, 1991* (Vol. 4, p. 121). Peeters Press.

[21] Chulak, I., Sutorius, V., and Sekerka, V. (1990). Benzothiazole compound XXXV. Synthesis of 3-substituted 2-benzylbenzothiazolium Salts and their Growth-regulating Effect on Triticum Aestivum L. *Chem. Pap, 44,* 131-138.

[22] Lacova, M., Chovancova, J., Hýblová, O., and Varkonda, S. (1991). Synthesis and Pesticidal Activity of Acyl Derivatives of 4-chloro-2-aminobenzothiazole and the Products of their Reduction. *Chem. Pap, 45* (3), 411-418.

[23] Krall, R. L., Penry, J. K., White, B. G., Kupferberg, H. J., and Swinyard, E. A. (1978). Antiepileptic drug development: II. Anticonvulsant drug screening. *Epilepsia, 19* (4), 409-428.

[24] Gullick, W. J. (1991). Prevalence of aberrant expression of the epidermal growth factor receptor in human cancers. *British Medical Bulletin, 47* (1), 87-98.

[25] Yates, P. C., Mccall, C. J., and Stevens, M. F. (1991). Structural studies on benzothiazoles. Crystal and molecular structure of 5, 6-dimethoxy-2-(4-methoxyphenyl-benzothiazole and molecular orbital calculations on related compounds. *Tetrahedron, 47* (32), 6493-6502.

[26] Manjula, S. N., Noolvi, N. M., Parihar, K. V., Reddy, S. M., Ramani, V., Gadad, A. K., and Rao, C. M. (2009). Synthesis and antitumor activity of optically active thiourea and their 2-aminobenzothiazole derivatives: A novel class of anticancer agents. *European journal of medicinal chemistry, 44* (7), 2923-2929.

[27] Isobe, T., Fukuda, K., Tokunaga, T., Seki, H., Yamaguchi, K., and Ishikawa, T. (2000). Modified guanidines as potential chiral superbases. 2. Preparation of 1, 3-unsubstituted and 1-substituted 2-iminoimidazolidine derivatives and a related guanidine by the 2-chloro-1, 3-dimethylimidazolinium chloride-Induced cyclization of thioureas. *The Journal of organic chemistry*, *65* (23), 7774-7778.

[28] Domino, E. F., Unna, K. R., and Kerwin, J. (1952). Pharmacological properties of benzazoles I. Relationship between structure and paralyzing action. *Journal of Pharmacology and Experimental Therapeutics*, *105* (4), 486-497.

[29] Brantley, E., Trapani, V., Alley, M. C., Hose, C. D., Bradshaw, T. D., Stevens, M. F., and Stinson, S. F. (2004). Fluorinated 2-(4-amino-3-methylphenyl) benzothiazoles induce CYP1A1 expression, become metabolized, and bind to macromolecules in sensitive human cancer cells. *Drug metabolism and disposition*, *32* (12), 1392-1401.

[30] Bao, G., Du, B., Ma, Y., Zhao, M., Gong, P., and Zhai, X. (2016). Design, Synthesis and Antiproliferative Activity of Novel Benzothiazole Derivatives Conjugated with Semicarbazone Scaffold. *Medicinal Chemistry*, *12* (5), 489-498.

[31] Lindgren, E. B., de Brito, M. A., Vasconcelos, T. R., de Moraes, M. O., Montenegro, R. C., Yoneda, J. D., and Leal, K. Z. (2014). Synthesis and anticancer activity of (E)-2-benzothiazole hydrazones. *European journal of medicinal chemistry*, *86*, 12-16.

[32] Ma, J., Chen, D., Lu, K., Wang, L., Han, X., Zhao, Y., and Gong, P. (2014). Design, synthesis, and structure–activity relationships of novel benzothiazole derivatives bearing the ortho-hydroxy N-carbamoylhydrazone moiety as potent antitumor agents. *European journal of medicinal chemistry*, *86*, 257-269.

[33] Zhang, H. Z., Drewe, J., Tseng, B., Kasibhatla, S., and Cai, S. X. (2004). Discovery and SAR of indole-2-carboxylic acid benzylidene-hydrazides as a new series of potent apoptosis inducers using a cell-based HTS assay. *Bioorganic & medicinal chemistry*, *12*(13), 3649-3655.

[34] Can, N. Ö., Osmaniye, D., Levent, S., Sağlık, B. N., Inci, B., Ilgın, S., and Kaplancıklı, Z. A. (2017). Synthesis of new hydrazone derivatives for MAO enzymes inhibitory activity. *Molecules*, *22*(8), 1381.

[35] Ilgın, S., Osmaniye, D., Levent, S., Sağlık, B., Acar Çevik, U., Çavuşoğlu, B., and Kaplancıklı, Z. (2017). Design and Synthesis of New Benzothiazole Compounds as Selective hMAO-B Inhibitors. *Molecules*, *22* (12), 2187.

BIOGRAPHICAL SKETCH

Premlata Kumari

Affiliation: Applied Chemistry Department, S. V. National Institute of Technology, Surat, Gujarat, India.

Education: PhD

Business Address: Applied Chemistry Department, S. V. National Institute of Technology, Surat-395007, Gujarat, India.

Research and Professional Experience: 13 Years

Professional Appointments:

- Associate Professor, Former Head, Applied Chemistry Department, S. V. National Institute of Technology, Surat, Gujarat, India.
- Life Member of Indian Chemical Society (LM No: 7440)
- Life member of Indian Council of Chemist (LM No. 1384)
- Life Member of Medicinal and Aromatic Plant Association of India (MAPAI)(LM No. 363)
- Member of Society for Ethnopharmacology (SFE/19/I-1376)
- Indian Science Congress (L30791)

Honors: Outstanding Reviewer by Elsevier in 2014 and 2015.

Publications from the Last 3 Years:

1. Singh, Raghuraj, Premlata Kumari, and Satyanshu Kumar. "Nanotechnology for enhanced bioactivity of bioactive phytomolecules." In *Nutrient Delivery*, pp. 413-456. Academic Press, 2017.
2. Kureshi, Azazahemad A., Chirag Dholakiya, Tabaruk Hussain, Amit Mirgal, Siddhesh P. Salvi, Pritam C. Barua, Madhumita Talukdar, Beena Chekunnath, Ashish Kar, Thondiath John Zachariah, Premlata Kumari, Tushar Dhanani, Raghuraj Singh, Ponnuchamy Manivel, and Satyanshu Kumar. "Simultaneous identification and quantification of three biologically active xanthones in *Garcinia* species using a rapid UHPLC-PDA method." *Acta Chromatographica* (2019): 1-10.
3. Kureshi, Azazahemad A., Chirag Dholakiya, Tabaruk Hussain, Amit Mirgal, Siddhesh P. Salvi, Pritam C. Barua, Madhumita Talukdar, Beena Chekunnath, Ashish Kar, Thondiath John Zachariah, Premlata Kumari, Tushar Dhanani, Raghuraj Singh, Ponnuchamy Manivel, and Satyanshu Kumar. "Simultaneous Identification and Quantification of Three Xanthones and Two Polyisoprenylated Benzophenones in Eight Indian *Garcinia* Species Using a Validated UHPLC-PDA Method." *Journal of AOAC International* 102, no. 5 (2019): 1423-1434.
4. Kureshi, Azazahemad A., Tabaruk Hussain, Amit Mirgal, Siddhesh P. Salvi, Pritam C. Barua, Madhumita Talukdar, Beena Chekunnath, Ashish Kar, Thondiath John Zachariah, Satyanshu Kumar, Tushar Dhanani, Raghuraj Singh, Premlata Kumari. "Comparative evaluation of antioxidant properties of extracts of fruit rinds of *Garcinia* species by *in vitro* assays." *Indian Journal of Horticulture* 76, no. 2 (2019): 338-343.
5. Lakhotia, Sonia R., Mausumi Mukhopadhyay, and Premlata Kumari. "Iron oxide (FeO) nanoparticles embedded thin-film

nanocomposite nanofiltration (NF) membrane for water treatment." *Separation and Purification Technology* 211 (2019): 98-107.
6. Lakhotia, Sonia R., Mausumi Mukhopadhyay, and Premlata Kumari. "Surface-modified nanocomposite membranes." *Separation & Purification Reviews* 47, no. 4 (2018): 288-305.
7. Lakhotia, Sonia R., Mausumi Mukhopadhyay, and Premlata Kumari. "Cerium oxide nanoparticles embedded thin-film nanocomposite nanofiltration membrane for water treatment." *Scientific reports* 8, no. 1 (2018): 4976.
8. Patel, Divyesh, Premlata Kumari, and Navin B. Patel. "Synthesis and biological evaluation of coumarin based isoxazoles, pyrimidinthiones and pyrimidin-2-ones." *Arabian Journal of Chemistry* 10 (2017): S3990-S4001.

In: Benzothiazole
Editor: Atakan Heijstek

ISBN: 978-1-53617-548-6
© 2020 Nova Science Publishers, Inc.

Chapter 3

ESIPT INSPIRED BENZOTHIAZOLE FLUORESCENT MOLECULES

Vikas Patil[1],*, Bhavana V. Mohite[2], Satish V. Patil[3] and Sharad Patil[4]

[1]University Institute of Chemical Technology,
North Maharashtra University, Jalgaon, India
[2]Department of Microbiology,
Bajaj College of Science, Wardha, India
[3]School of Life Science,
North Maharashtra University, Jalgaon, India
[4]Department of Chemistry, SPDM, Arts, Commerce Science College,
Shirpur, India

ABSTRACT

Excited state intramolecular proton transfer (ESIPT) based 2-substituted benzothiazole fluorescent moleculeshavegainedconsiderable attention in the pastfew yearsas ausefulmolecule in high-tech and classical application. It was due to its desirable unique photo-physical properties induced due to the proton transfer in an excited state. The photo-physical

---

* Corresponding Author's E-mail: vikasudct@gmail.com.

properties of these benzothiazole ESIPT derivatives make them an interesting moiety and were studied as a function of pH and viscosity. High fluorescence quantum efficiencies and photo-stability in the micro-environment give enhancement in fluorescence. The observations are bound to the movement of the molecules in the solvent as non-bonding interactions with the surrounding environment in solution. It was also relevant to find out the relative fluorescence quantum yields of these derivatives by using secondary reference standards such as anthracene and fluorescence. The development of ESIPT chromophore with high fluorescence quantum efficiencies and a long fluorescence lifetime in the solid state for benzothiazoles is always a challenging issue due to the unpredictable mechanism of fluorescence in the solid state. Here a report on design strategies, detailed photo-physical properties, and their applications will help in assisting researchers to overcome existing challenges in designing novel ESIPT benzothiazole fluorescent molecules for promising applications. Present chapter deals with the developments of fluorescent benzothiazole ESIPT molecules with a focus on fluorescence properties.

Keywords: Benzothiazole, ESIPT, Fluorescence, Photo-physical, Synthesis

INTRODUCTION

Hantzsch and Waber first described thiazole in 1887 and its structure was confirmed by Popp in 1889. Thiazole is a five-member ring heterocyclic compound consisting of nitrogen and a sulfur atom in the ring. Benzothiazole is an organic compound bearing a heterocyclic nucleus (thiazole) that imparts a broad spectrum of biological and high-tech applications. The basic structure of benzothiazole is the combination of a benzene ring fused with 4, 5 positions of thiazole. Benzothiazole is a colorless, slightly viscous liquid with a melting point of 2°C and a boiling point of 227-228°C. The density of benzothiazole is 1.238g/ml (25°C). Benzothiazole has no household use. It is usedforindustry and research work purpose which isvery beneficial for the development of the various pharmaceutical and biological active compounds. Classically benzothiazoles has been prepared by different way (Scheme 1).

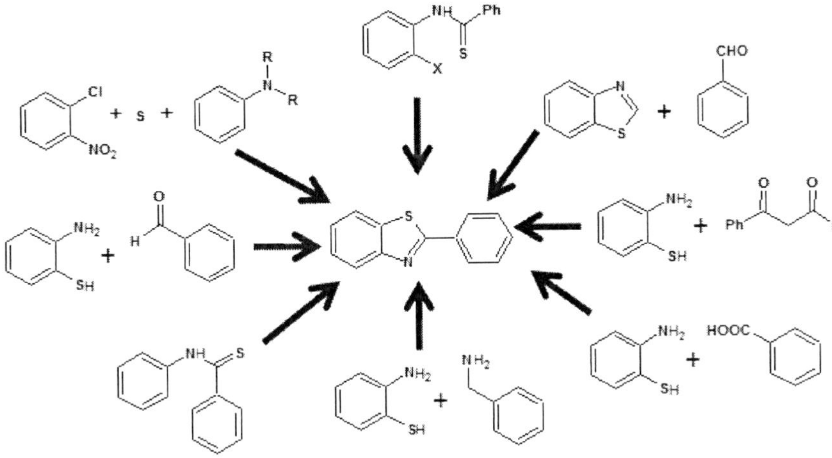

Scheme 1. The different way of synthesis of benzothiazole combining different chemical species.

Few catalytic reactions are also promoted for the specific benzothiazole preparations.

BENZOTHIAZOLE FLUORESCENT

Benzothiazole molecules do not have inbuilt structural ability to have the fluorescence emission after photo-excitation by UV-visible light. The molecular structure of benzothiazole does not allow establishing the HOMO and LUMO energy levels even by the addition of electron donating and withdrawing mechanisms. The fluorescence may be inducing in benzothiazole by the mechanism of excited state intramolecular proton transfer (ESIPT) phenomenon. The fundamental requirements of the ESIPT process are the presence of the accessible acidic hydrogen, basic element (*N, S, O*) able to accommodate hydrogen and the suitable geometry of the molecular system. ESIPT inspired benzothiazole mostly requires accessible acidic *ortho* substituted protons *(–OH or -NH₂)* at 2-position and basic centres =*N* (1) hydroxy benzothiazole (HBT). Within this basic =*N*- we can induce Sulphur (*S*) heteroatom as benzothiazole. As such, there is no role of sulphur in the phenomenon of fluorescence. Still, a variety of molecules, including benzothiazole [1-4], which capable of showing ESIPT phenomenon, are reported in the literature, and their photophysical properties were also studied as a function of the micro-environment. It is the commonly observed large Stokes shift of benzimidazole, benzothiazole, benzoxazole, and benzotriazole units within the ESIPT system (Figure 1) [5-7]. The ESIPT and fluorescence properties in benzothiazole molecules give interesting applications [8].

ortho - **hydroxy benzothiazole (HBT) (1)**

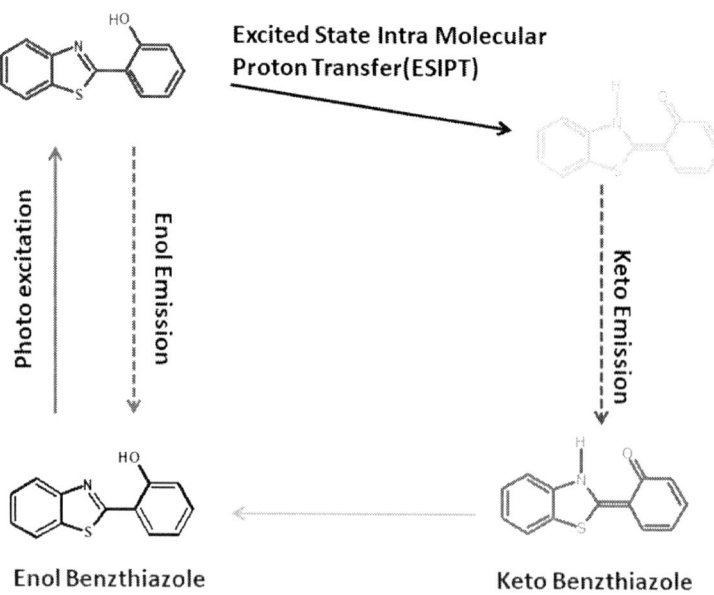

Figure 1. Mechanism of Fluorescence in benzothiazole by ESIPT.

ESIPT Inspired Benzothiazole Fluorescent Molecules

[Structures shown:
- 2-(Triazole-5-naphthalene sulfonate) benzothiazole (2)
- 4-(*N,N*-dialkyl amine) benzothiazole (3)
- 2-(Dimethyl quinoline sulfonate) benzothiazole (4)
- 2-(Diazo 4-dialkylaminobenzene) benzothiazole (5)]

Earlier in 1955, Arthur and co-workers have been reported the 2-(triazole-5-naphthalene) substituted benzothiazole molecules as fluorescent brightening agents (2)[9] where it was again modified to para *N, N-dialkyl amine s*ubstituted benzothiazole (3)[10]. The sulphonic acid group was typically induced to impart water solubility while using it as a fluorescent brightening agent [9, 10]. Further, the 2-(dimethyl quinoline sulfonate) substituted benzothiazole (4) fluorescent brightening agent was also evaluated as a fluorescent brightening activity[11]. Apart from this, a diazotization was also performed at the second position of benzothiazolefor dyeing keratinic fibers, especially for human hair (5)[12].Wang and co-workers synthesized the different benzothiazole an ESIPT 2-(2'-hydroxyphenyl) benzothiazole coloured organic compounds [13]. The colour characteristics were dependent on the different substituent's at 5-positions of phenyl ring substituted to benzothiazole moiety. The chromophoric nature was driven by the electronic push-pull character at 5th and 4th position of attached phenyl ring given in Table 1compounds6 to 11. In the scheme 1 benzothiazole unit acts as an electron deficient center to accept the electron push from amine functional groups attached to phenyl substituent. The major influence on the fluorescence characteristics is substituent groups, which driven the emission wavelength as well as the fluorescence quantum yield. As shown in Table 1, carbazole substituent at the para position (6) has an emission at 502nm while the quantum yield is 91.68%. Similarly, the carbazole substituent at *meta* position (7) has an emission at 532nm, while its quantum yield is lower to 66.86%. For the N,

N-dimethyl substituent at *para*(8) and *meta*(9) position, the fluorescence emission is at 443nm and 630nm, respectively while, *para* and *meta* substituent has fluorescence quantum yield 29.07% and 5.65% respectively. Rather in the case of N N diethyl substituent's emission at *para* and *meta* is 461nm and 600nm respectively while quantum yield is 30.16% and 37.54% respectively. The observation is clear that *meta* substituent's have a long wavelength emission than *para* substituent's. However, there is no descending and ascending trend in the fluorescence quantum yield. It is the intrinsic property of fluorescent material, which is independent of the position of the substituent group. Further, it was convicted from single crystal structure that the phenyl substituent groups at phenyl ring of benzothiazole units orient more planner nature in terms of achieving easy intramolecular hydrogen transfer i.e., keto-enol tautomerism. It is even substitution at rare position adjusted in such a way that orient the benzothiazole ring and phenyl ring orient dominating six-member keto-enol planer structure. As observed in *para* (10) and *meta* (11), substituted phenyl ring gives more quantum yield as compared to N, N di-substituted compounds. This gives majorly solid state fluorescence as well as reported for high fluorescence quantum yield 92% (Table 1)(6), which is very rare for substituted benzothiazole. Creating a small dihedral angle favours the intra-molecular hydrogen transfer giving non-radiative pathways induces molecule fluorescent even in solid state. These properties of solid state emission were utilized in fabricating the electroluminescent (EL) device where 4,4-Bis[N-(1-naphthyl)-N-phenyl-amino]biphenyl (NPB) and 1,3,5-tri- (1-phenyl-1H-benzo[d]imidazol-2-yl)phenyl (TPBI) were used as hole-transporting material and electron-transporting-hole-blocking material respectively. Among all these benzothiazole substituent's *meta*-carbazole substituted benzothiazole gives good EL characteristics with a layer of [ITO/NPD (15nm)/ *meta*-carbazole benzothiazole (5nm)/TPBI (75nm)/LiF (0.5nm)/Al (200nm)][13]. A similar study was also reported for the compounds (26) coumarin substituted benzothiazole, where coumarin itself was fluorescence unit and combined two different fluorescents units has significant fluorescence quantum efficiency of 49% [14].

Table 1. Different phenyl substituted benzothiazole and their relation between emission and fluorescence quantum yield

para-carbazole
(6) λ_{em} = 502 nm, ϕ_{FL} = 91.68%

meta-carbazole
(7) λ_{em} = 532 nm, ϕ_{FL} = 66.86%

para-N, N- Methyl
(8) λ_{em} = 443 nm, ϕ_{FL} = 29.07%

meta-N, N Methyl
(9) λ_{em} = 630 nm, ϕ_{FL} = 5.65%

para-N, N- diphenyl
(10) λ_{em} = 461 nm, ϕ_{FL} = 30.16%

meta-N, N- diphenyl
(11) λ_{em} = 600 nm, ϕ_{FL} = 37.54%

2,4- Dibenzoyl phenols
(12)(2,4 DBPT) λ_{em} = 570 nm, ϕ_{FL} = 68%

2,6- Dibenzoyl phenols
(13)(2, 6 DBPT) λ_{em} = 576, ϕ_{FL} = 62%

Table 1. (Continued)

2,6 di-benzthiazole, 4 alkoxyphenol (2, 6 DBT4AP), R = -OCH₃, -OC₂H₅, -OC₃H₇, -OC₄H₉ **(14)** OMe: λ_{em} = 647 nm, ϕ_{FL} = 17%, OEt: λ_{em} = 633 nm, ϕ_{FL} = 32%, OPr: λ_{em} = 589 nm, ϕ_{FL} = 38%, OBt: λ_{em} = 592 nm, ϕ_{FL} = 44%	R: -methyl, -hexyl, -octyl, -dodecyl **(15)** λ_{em} = 405–552 nm, ϕ_{FL} = 54–68% (solid state), 1-11% (in solvent)
TAA **(16)** λ_{em} = 552 nm, ϕ_{FL} = 78.1%	TA: R = -H, -tBu **(17)**
DHIA **(18)** λ_{em} = 393 nm (enol) 509 nm (keto), ϕ_{FL} = 0.2% Reference 91	1a-h **(19)** 2-(2,2'-bithiophen-5-yl)-1,3-benzothiazole | | | ϕ | |---|---|---| | a | R= H | 0.25 | | b | R= OMe | 0.58 | | c | R= OEt | 0.56 | | d | R= NMe₂ | 0.48 | | e | R= NEt₂ | 0.46 | | f | R= N(Pr-i)₂ | 0.41 | | g | R= piperidino | 0.52 | | h | R= morpholino | 0.48 |

ESIPT Inspired Benzothiazole Fluorescent Molecules

	λ_{em} [nm]
	454
	498
	500
	587
	587
	588
	593
	582
(20) 2-(10 H phenothiazine)-2-yl-1,3-benzothiazole	(21)
(22)	(23) benzothiazole based fluorescein
(24)	(25) λ_{em} = 443-530 nm, ϕ_{FL} = 5.65-91.68%
(26) λ_{em} = 363 and 514 nm, ϕ_{FL} = 49% Reference 127	(27) λ_{em} = 500, 550 and 650 nm, ϕ_{FL} = 5-11% Reference 129

Table 1. (Continued)

(28)	(29)
(30)	(31)

While excess addition of phenyl ring in hydroxyl benzothiazole does not have a significant effect on fluorescence [15-16].

Sakai and et al. have reported the di-benzothiazolyl phenols (2,4-DBTP) (12) and 2,6-dibenzothiazolylphenols (2,6-DBTP) (13) which are blue-white-yellow light emitting compounds (Table 2). It gives the solvent dependent fluorescence emission in the solid state while among the solvents, chloroform gives blue shifted emission with a significant decrease in fluorescence quantum yield for 2, 4 DBTP (λ_{em} = 554nm, ϕ_{FL} = 9%) [17]. The properties of this 2,4-DBTP and 2,6-DBTP are highly distinct from classical benthiazoles. Three different emissions were created due to the anionic, keto, and enol forms (Scheme 2). The identical emission band at 570, 576, and 554nm was observed for both DBTP derivatives. Due to this excess emission 2,4-DBTP covers the entire visible light region for fluorescence in solvent mixture DMF:CHCl$_3$ (1: 1). The extent of fluorescence quantum yield was dependent on the strong intramolecular hydrogen bonding [18] which was envisaged by the synthesizing difference alkoxy substituted 2, 6 DBTP derivatives. The trend of fluorescence quantum yield was increasing with increasing electron donating nature of substituent given as OMe: ϕ_{FL} = 17%, OEt: ϕ_{FL} = 32%, OPr: ϕ_{FL} = 38%, OBt: ϕ_{FL} = 44% (Table 1)(14). Even in the 2-(aminophenyl) benzothiazole,

ESIPT Inspired Benzothiazole Fluorescent Molecules 109

a cationic anionic mechanism was observed promoting the fluorescence emission of the single molecules (Scheme 3). It is correlated with the ionizing capacity of molecules in a solvent, which is truly controlled by the pKa value in solvent dissociation [19]. The pKa value is responsible for the dissociation in the different solvents which gives the response to the fluorescence emission which was truly controlled by the solvents and its mixture where it continuously changes the dipole moment [20].Similar to the anionic structure of 2, 6-DBTP.

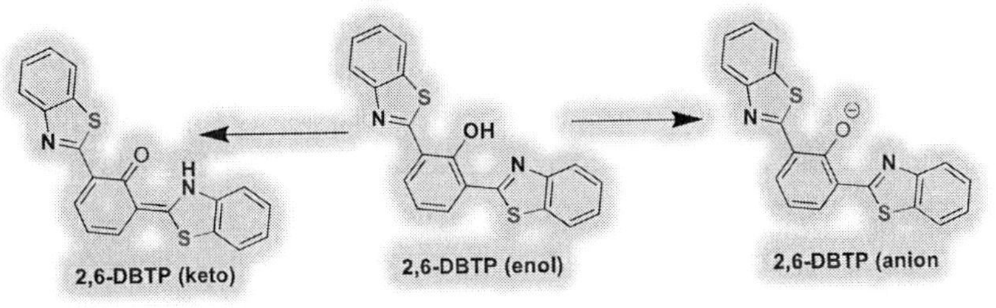

Scheme 2. Anionic, keto and enol forms of 2, 6 DBTP.

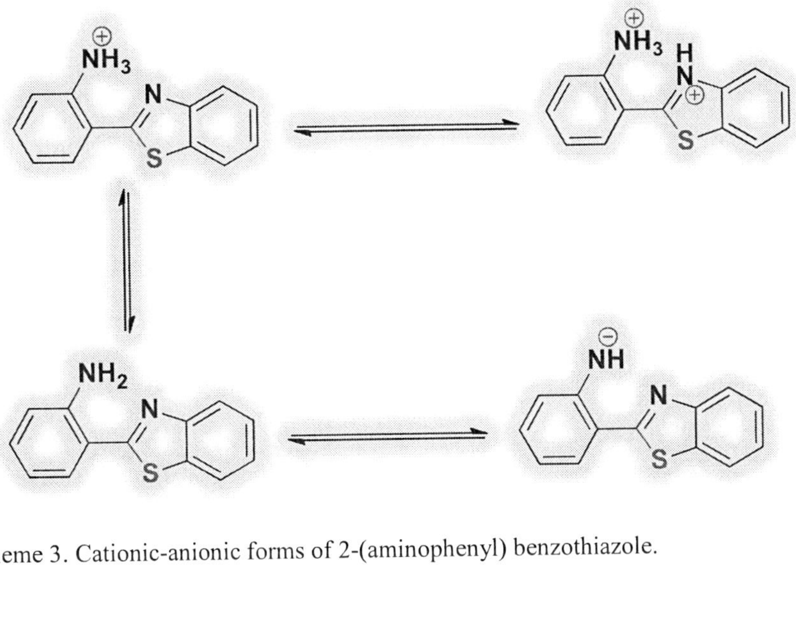

Scheme 3. Cationic-anionic forms of 2-(aminophenyl) benzothiazole.

Padalkar et al. synthesized a new range of hydroxy benzothiazoleunits substituted to 7, 7' position of benzcarbazole with substituent's at and alkyl substituents at 9, 9' position. They show absorption in the region of 325-344nm, while emission is at 405-552nm (15). The distinct two emissions were observed in the solvent, while single emission was observed in solid state. The two emissions in solvents are corresponding for the keto and enol tautomerism. Fluorescence quantum yield was observed much higher in the solid state (ϕ_{FL} = 54–68%), which is greater than the in the solvent (ϕ_{FL} = 1–11%) (15). [21] Even the cis and trans geometry of the substituted benzothiazole compounds N-(3-(benzo[d]thiazol-2-yl)-4-hydroxyphenyl) anthracene-9-carboxamide (TAA) (16) and N-(3-(benzo[d]thiazol-2-yl)-4-hydroxyphenyl)- phenyl-carboxamide (TA)(17) has also prominent effect on the fluorescence spectra of the compounds (Table 1). As the di-substitution of N-(3-(benzo[d]thiazol-2-yl)-4-hydroxyphenyl) at 1, 3 position in TAA (DHIA) (18) deactivate the fluorescence effect and gives λ_{em} = 509nm, ϕ_{FL} = 0.2%. Also, deactivation of ESIPT was found in keto form due to the cis-trans isomerization and it shows quenching of fluorescence in solution. The tight molecular packing effectively blocks the cis-trans isomerization, which significantly gives the blocking of deactivation and generated radiative emission [22, 23].

The substitution of different heterocycles at 2-position of benzothiazole majorly affect the fluorescent properties of the compounds. In 2-thiophen (19) and 2- (10 H phenothiazine) (20) substituted 1,3-benzothiazole affect the quantum yields of different substituent [24]. The effect of the substituent is detail studied in thiophene substituent (19) and found some relation where quantum yield is from 58% - 25% while emission is from 450-600 nm. The electron donating substituent's were promoting quantum yield as methoxy substituent gives high quantum yield of 58% while unsubstituted found less quantum yield of 25% as summarised in Table 1[25].

An extension of bridging with the double bond induces the styryl functional group to the 2-position of benzimidazole, which was tested as fluorescent whitening agents and no loss of fluorescence even after the extension of chain at the 2-position of benzothiazole (21) [26]. A blue fluorescent emissive series of compounds benzthiazoyl pyrazoline (22) was synthesized, having emission at wavelengths 450nm. It was observed that the functional group at 3 position of pyrazoline has a strong influence on the emission (22)[27]. A series of new ESIPT benzothiazole fluorescein was designed and synthesized from different anhydrides. The photo-physical properties of these ESIPT benzothiazole fluorescein compounds were investigated as a function of pH and viscosity. Enhancement in fluorescence emission was observed with an increasing percentage of glycerine in water. The structural geometry was also optimized by using density functional (DFT) theory computational calculations. It was observed distinct two emissions and determined the two different quantum yield corresponding to the two emissions of fluorescein based benzothiazole ESIPT compounds [28]. ESIPT benzothiazole fluorescein has potent applications as pH and viscosity sensor. Rather than ESIPT a push a pull centre is created in (Z)-2-(2- Phenyl-4H-benzo[4,5]thiazolo[3,2-a]pyrimidin-4-ylidene)acetonitrile derivatives (24) [29] generating fluorescence emission. The detail about the solvatochromic study and optimization by DFT was also reported [9].

Similarly, 2-(2′,2′′-bithienyl)-1,3-benzothiazoles was synthesized and find applications as additives in textile dyeing, plastics, as tunable laser dye, microbialstains, fluorescent markers, materials science and opto-electronics[30].

APPLICATION OF BENZOTHIAZOLE FLUORESCENT COMPOUNDS

Fluorescent compounds have found widespread use in scientific and industrial areas, for example, as fluorescent brightening agents for textiles, plastics, inks and paints, tunable laser dyes lasers, and biological stains. Other applications include electroluminescent and liquid crystal displays, solar collectors, materials science, and optoelectronics. The benzothiazole nucleus appears in many fluorescent compounds that have high tech applications as a result of the ease of synthesis of this heterocycle and the high fluorescence quantum yields obtained when this small, rigid moiety is present in compounds.

benzothiazole is one of the essential heterocyclic compounds, a weak base, having varied biological activities, and still of great scientific interest nowadays. They have wide applications in bio-organic and medicinal chemistry fields such as drug discovery.benzothiazole is an honored bicyclic ring system. The significant biological activities resulted in great pharmaceutical applications, and therefore its synthesis is of considerable interest. benzothiazole moieties are part of compounds showing numerous biological activities such as anti-tubercular, antimicrobial, antimalarial, anticonvulsant, anthelmintic, anti-inflammatory, anti-tumor, anti-diabetic, analgesic, central muscle relaxant activities and also work against the neurodegenerative disorders and local brain ischemia. They have also found applications in industry as antioxidants, vulcanization accelerators. benzothiazole, such as 2-aryl benzothiazole, received much attention due to the unique structure and its uses as radioactive amyloid imagining agents and anticancer agents.

Furthermore, it includes marine or terrestrial natural compounds that have tremendous biological activities. Chen et al. developed a highly specific and ratiometric fluorescent probe (28) for the detection of hypochlorite in living cells. The probe (28) exhibits rapid response and high selectivity to hypochlorite over other reactive oxygen species (ROS) and reactive nitrogen species (RNS), accompanied the shifted fluorescence from yellow to the blue [31].

Han et al.has been designed and developed to a single 2-(2'-Hydroxyphenyl) benzothiazole (HBT) (29) derivative achieved pure white-light emission composed of enol, keto, as well as phenolic anion emission. Based on the interactions of an electron-donating/-withdrawing framework with solvents involved the intermolecular hydrogen bond, the excited-state intramolecular proton transfer (ESIPT) process, as well as deprotonation, a single HBT derivative achieves white-light emission with CIE coordinates of (0.33, 0.33) under mild conditions in binary solvents (Ethanol:Acetonitrile).

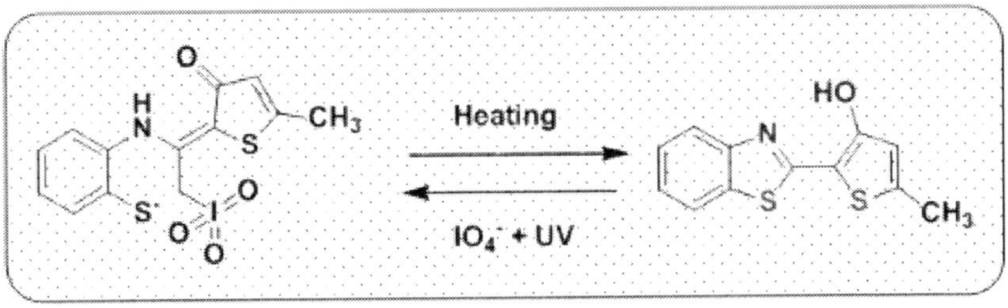

Scheme 4. Mechanism of decrease in fluorescence sensitized by selective addition of IO4.

This work presents a novelway for the rational design of an ESIPT molecule to achieve white-light generation under mild conditions [32].

Ma et al. developed a selective turn-on fluorescent probe (31) that was synthesized by modulation of the excited-state intramolecular proton transfer (ESIPT) process of 2-(2'-hydroxy-4'-diethylaminophenyl) benzothiazole for detection of biological thiols (31). The excellent properties of it make it useful for monitoring thiols in living cells as well as a suitable fluorescent probe for the specific detection of thiols [33].

Hong et al. designed and developed a hydroxythiophene-bearing benzothiazole (HB) in phosphate buffer solution showed strong yellow fluorescence emission, which selectively and sensitively decreased by addition of IO_4^-. The reaction between hydroxy-benzothiazole and IO4- was dramatically accelerated by UV irradiation. The fluorescence emission of hydroxy-benzothiazole was gradually recovered upon thermal heating of the IO_4^--adducted product, indicating IO_4^- release. Using these properties, the solution containing HB and IO_4^- has been used as a fluorescence-based security ink [34, 35].

CONCLUSION

In this chapter, we have discussed in brief about some commonly developed benzothiazole derivatives and various structural alterations conducted on benzothiazole ring and preferential specificities imparted in their sensing responses. The majorly existing fluorescent molecules as an

ESIPT benzothiazole are discussed and their few applications. The small and simple benzothiazole nucleus if present in compounds involved in research aimed at evaluating new products that possess interesting valuable high technological and classical applications. The chemistry and biological study of heterocyclic compounds have been an exciting field for a long time in the medicinal chemistry of benzothiazole derivative.

REFERENCES

[1] Sinha, H., Dogra, S. (1986). *Chem. Phys.,* 102:337 - 347.
[2] Das, K., Mjumder, D., Bhattacharyya, K. (1992). *Chem. Phys. Lett.,* 198:443 - 448.
[3] Douhal, A., Amat-Guerri, F., Lillo, M., Acuna, A. (1994). *J. Photochem. Photobiol. A: Chem.,* 78:127 - 138.
[4] Rios, M. *J. Phys. Chem.,* (1995).99: 12456 - 12460.
[5] Das, K., Sarkar, N., Ghosh, A., Majumdar, D., Nath, D., Bhattacharya, K. (1994). *J. Phys. Chem.,* 98:9126 - 9132.
[6] Kelly, R., Schulman, S., Schulman, S. (1988). *Molecular luminisons spectroscopy, methods and applications Part −2,* Willy Interscience New York, Chapter 6.
[7] Ghiggino, K., Scully, A., Leaver, I. (1986). *J. Phys. Chem.,* 90:5089 - 5093.
[8] Padalkar,Vikas S., Abhinav Tathe, Gupta, Vinod D., Patil, Vikas S., Kiran Phatangare, Sekar, N. *J. Fluoresc.,* (2012) 22:311 - 322.
[9] Baum, Arthur A. and Sanders, Paul A. (1955). US2,713,054.
[10] Zwigmeyer Fritjof, Edgewood Hills (1955). US 2,715,629.
[11] Clarke, Ray Allen, Pitman, N.J. (1964). US 3,152,132.
[12] Kalopissis Gregoire, Bugaut Andree (1971). US 3,578,386.
[13] Yao, D., Zhao, S., Guo, J., Zhang, Z., Zhang, H., Liu,Y. and Wang, Y. *J. Mater. Chem.,* 2011, 21, 3568 - 3570.
[14] Xie, L., Chen, Y., Wu, W., Guo, H., Zhao, J. and Yu, X. *Dyes Pigm.,* 2012, 92, 1361 - 1369.

[15] Cui, L.,W. Zhu, Y. Xu and X. Qian, *Anal. Chim. Acta,* 2013. 786, 139 - 145.

[16] Sakai, K., H. Kawamura, N. Kobayashi, T. Ishikawa, C. Ikeda, T. Kikuchi and T. Akutagawa, *Cryst. Eng. Comm.,* 2014. 16, 3180 - 3185.

[17] Sakai, K.,T. Ishikawa and T. Akutagawa, *J. Mater. Chem. C,* 2013. 1, 7866 - 7871.

[18] Sakai, K.-I., S. Takahashi, A. Kobayashi, T. Akutagawa, T. Nakamura, M. Dosen, M. Kato and U. Nagashima. *Dalton Trans.,* (2010) 39, 1989 - 1995.

[19] Joy Krishna Dey, Dogra, Sneh K. (1991). 64, 3142-3152.

[20] Dogra, S.K. *Journal of Photochemistry and Photobiology A: Chemistry*, (2005). 172 185 - 195.

[21] Padalkar, V. S., D. Sakamaki, N. Tohnai, T. Akutagawa, K. Sakai and S. Seki, *RSC Adv.,* 2015, 5, 80283 - 80296.

[22] Cai, M., Z. Gao, X. Zhou, X. Wang, S. Chen, Y. Zhao, Y. Qian, N. Shi, B. Mi, L. Xie and W. Huang, *Phys. Chem. Chem. Phys.,* 2012, 14, 5289 - 5296.

[23] Qian, Y., S. Li, G. Zhang, Q. Wang, S. Wang, H. Xu, C. Li, Y. Li and G. Yang, *J. Phys. Chem. B,* 2007, 111, 5861 - 5868.

[24] Costa,Susana P. G., Ferreira, A, Joao A., b Gilbert Kirschc and Oliveira-Campos, Ana M. F. *J. Chem. Research (S)*, (1997) 314 - 315.

[25] Batista, Rosa M. F.,Costa Susana P. G. and M. Manuela M. Raposo. *Tetrahedron Letters,* 45 (2004) 2825 - 2828.

[26] Belgodere, E.,R. Bossio, S. Chimichi, V. Parrinit and R. Pepino. *Dyes and pigmetes*, 4 (1983) 59-71.

[27] Ji Shun-Jun, Shi Hai-Bin. *Dyes and Pigments,* 70 (2006) 246-250.

[28] Patil, Vikas S., Padalkar, Vikas S., Abhinav B. Tathe, N. Sekar, *Dyes and Pigments,* 98 (2013) 507-517.

[29] Patil, R.S., Patil, A.S., Patil, V.S., Jirimali, H.D., Mahulikar, P.P. *J. Lumin.,* 210 (2019) 303-310.

[30] Batista, Rosa M. F.,Costa, Susana P. G. and M. Manuela M. Raposo (2004). *Tetrahedron Letters,* 45, (13), 2825-2828.

[31] Chengcheng Chang, Fang Wang, Jian Qiang, Zhijie Zhang, Yahui Chen, Wei Zhang, YongWang, Xiaoqiang Chen. *Sensors and Actuators B: Chemical,* DOI 10.1016/j.snb.2016.11.123.

[32] Jinling Cheng, Di Liu, Lijun Bao, Kai Xu, Yang Yang and Keli Han; *Chemistry an Asean Journal,* 10.1002/asia.201402779.

[33] Feng, S., Li, X., Ma, Q., Liang, B. and Ma, Z. *Anal. Methods,* 2016, DOI: 10.1039/C6AY02140A.

[34] Hong, K.-I., Choi, W. H., Jang, W.-D. *Dyes and Pigments,* (2018). doi: 10.1016/j.dyepig.2018.11.026.

[35] Satam, M. A., Telore, R. D., Tathe, A. B., Gupta, V. D., Sekar, N. *Spectrochim. Acta, Part A,* 2014, 127, 16 - 24.

In: Benzothiazole
Editor: Atakan Heijstek

ISBN: 978-1-53617-548-6
© 2020 Nova Science Publishers, Inc.

Chapter 4

BENZOTHIAZOLE: A PROMISING SCAFFOLD FOR THE DEVELOPMENT OF ANTICANCER AGENTS

Shah Alam Khan[1,], Asif Husain[2], Yaseen Moosa Al Lawatia and Saif Ahmad[3]*

[1]College of Pharmacy, National University of Science and Technology, Muscat, Sultanate of Oman
[2]Department of Pharmaceutical Chemistry, School of Pharmaceutical Education and Research, Jamia Hamdard, New Delhi, India
[3]Department of Neurosurgery, Barrow Neurological Institute, St. Joseph's Hospital and Medical Center (SJHMC), Dignity Health, Phoenix, AZ, US

ABSTRACT

Benzothiazole is an example of bicyclic ring system in which benzene and thiazole, a five membered heterocyclic ring with two hetero atoms *viz.* nitrogen and sulfur, are fused together. This versatile biologically active scaffold has been found to be present in various natural products of therapeutic importance.

---

[*] Corresponding Author's E-mail: shahalam@nu.edu.om; sakhan@omc.edu.om.

Literature review indicates that benzothiazole nucleus has played an important role in drug design and drug discovery process of newer drugs. Indeed, it serves as a core nucleus in few clinically useful drugs such as viozan, riluzole, ethoxazolamide, etc. Several other broad spectrum molecules developed based on this promising scaffold are currently under different phases of clinical trials. In lieu of its diverse pharmacological activities, simple structure, ease of synthesis and substitution with different functionalities, medicinal chemists consider this skeleton as a privileged scaffold for the development of safe and efficacious chemotherapeutic agents. This review provides recent updates on the usefulness, structure activity relationship and potential of benzothiazole derivatives for the development of novel anticancer agents.

Keywords: anticancer, benzothiazole, drug design, heterocyclic

ABBREVIATIONS

ABTS	2,2'-azino-bis(3-ethylbenzothiazoline-6-sulfonic acid
AD	Alzheimer's Disease
BCL2	β-cell lymphoma 2
CDK	Cyclin-dependent kinase
COPD	Chronic obstructive pulmonary disease
CYPp450	Cytochrome p450
DPPH	Diphenyl picryl hydrazyl
ERK	Extracellular signal-regulated kinase
FDA	Food and drug administration
FITC assay	Fluorescein isothiocyanate assay
GPx	Glutathione peroxidase
GSH	Reduced glutathione
HDAC	Histone deacetylase
hERG	Human Ether-à-go-go-Related Gene
hGC-ALP	Human germ cell alkaline phosphatase

HGFR	Hepatocyte growth factor receptor
Hsp90	Heat shock protein 90
LPAAT-β	Lysophosphatidic acid acyl transferase-β
MAPK	Mitogen-activated protein kinase
MBTU	Methabenzthiazuron
mTOR	Mammalian target of rapamycin
MTT	3-(4,5-dimethylthiazol-2-yl)-2,5-diphenyltetrazolium bromide
NEFs	Nucleotide-exchange factors
nM	nanomolar
NSCLC	Non-small cell lung cancer
PI3K	Phosphoinositide-3-kinase
PPARs	Peroxisome Proliferator-Activated Receptors
ROS	Reactive oxygen species
SAR	Structure activity relationship
SOD	Superoxide dismutase
SRB	Sulforhodamine B
TAC	Total antioxidant capacity
TKI	Tyrosine kinase inhibitors
VEGFR	Vascular endothelial growth factor receptor

1. INTRODUCTION

Heterocyclic compounds have played a major role in the drug discovery. In fact, most of the bioactive natural products and commercially available drug molecules contain a heterocyclic ring in their structure support the hypothesis that their presence in the molecule has significant effect on bioactivity (Katritzky & Rees, 1984).

These heterocyclic compounds are present in numerous biomolecules, xenobiotics, phytochemicals etc., and being the core structural feature of large number of drugs are responsible for their pharmacological actions. Heterocyclic compounds containing both nitrogen and sulfur atoms comprise an interesting class of natural and synthetic compounds owing to their broad spectrum of activities.

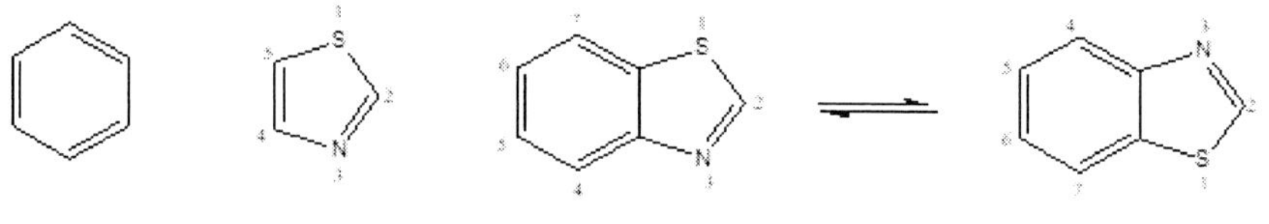

Figure 1. Chemical structure of benzene, thiazole and tautomeric forms of benzothiazole.

Table 1. Physical properties of 1,3 benzothiazole compound

S. No	Physical properties	
1.	IUPAC Name	1,3-Benzothiazole
2.	Molecular Formula	C_7H_5NS
3.	Composition	C (62.19%), H (3.73%), N (10.36%), S (23.72%)
4.	Molecular Weight	136.19
5.	Boiling Point	227-228°C
6.	Melting Point	2°C
7.	Density	1.244 g/cm^3
8.	Physical appearance	Colorless, slightly viscous liquid
9.	Solubility	Sparingly soluble in water, soluble in alcohol and carbon disufhide, highly soluble in ether and acetone
10.	pKa	1.2 (weak base)
11.	Index of Refraction	1.689 ± 0.02
12.	UV (EtOH)	λ_{max} 217, 251, 285, 295

Benzothiazole (C_7H_5NS) is an example of bicyclic planar ring system in which benzene is fused with a five membered 'thiazole' heterocycle containing two hetero atoms *viz.* sulfur and nitrogen at 1 and 3 positions. The 1,3 benzothiazole moiety is very stable and shows two tautomeric forms. It is a colorless and slightly viscous liquid at room temperature which is highly soluble in acetone and ether. The physical properties of 1,3 benzothiazole is given in Table 1.

1.1. Benzothiazole Containing Therapeutic and Diagnostic Agents

Benzothiazole and its derivatives form an important class of therapeutic medicinal agents. This pharmacophore has been associated with diverse pharmacological and biological actions viz. anticancer, antimicrobial, immunosuppressant, antiviral, anticonvulsant, antidiabetic, antidepressant, antitubercular, analgesic & anti-inflammatory, antioxidant, diuretic, antihelmintic and neuroprotective activities. In lieu of its diverse pharmacological activities, simple structure, ease of synthesis and substitution with different functionalities, medicinal chemists consider this skeleton as a privileged scaffold for the synthesis and development of new therapeutic agents. Literature review indicates that benzothiazole nucleus has played a pivotal role in drug design and drug discovery process of several newer drugs some of which are used clinically. The chemical structure and use of drugs containing benzothiazole nucleus is presented in Table 2.

Table 2. Chemical structure and uses of therapeutic agents containing benzothiazole nucleus

Comp No.	Name of the drug	Chemical structure	Biological action/use	Reference
1	Bentaluron		Fungicide	Khokra et al., 2019
2	Dimazole		Antifungal	Morton, 1960
3	Dithiazanine Iodide		Anthelmintic	Ding & Sutlive, 1960
4	Ethoxazolamide		Carbonic anhydrase inhibitor, glaucoma, diuretic, parkinson disease	Supuran, 2008
5	Flutemetamole		β-amyloid PET imaging agent	Nelissen et al., 2009

Comp No.	Name of the drug	Chemical structure	Biological action/use	Reference
6	Frentizole		Antiviral, Immunosuppressant, used in rheumatoid arthritis and systemic lupus erythematosus	Scheetz et al., 1977.
7	Halethazole		Antiseptic agent	Shi-Wei et al., 2005.
8	Methabenzthiazuron (MBTU)		Herbicide	Hartley, 1987
9	Perospirone		Antipsychotic agent	Arakawa et al., 2010
10	Pittsburg compound B		Amyloid imaging agent in Alzheimer's Disease	Klunk et al., 2004

Table 2. (Continued)

Comp No.	Name of the drug	Chemical structure	Biological action/use	Reference
11	Phortress (NSC 710305)		Antitumor prodrug	Fichtner et al., 2004
12	PMX 610		Antitumor	Aiello et al., 2008
13	Pramipexole		Anti Parkinsons disease, restless legs syndrome, Anti AD	Vinodhini et al., 2011.
14	Probenazole		Herbicide	Reemtsma et al., 1995
15	Revospirone		Anxiolytic agent	Jain et al., 2010

Comp No.	Name of the drug	Chemical structure	Biological action/use	Reference
16	Riluzole		Glutamate receptor antagonist, to treat amyotrophic lateral sclerosis, anti-depressant and anxiolytic	Satyanarayana et al., 2008
17	Sagopilone (ZK-EPO)		Epothilone B analog microtubule-stabilizing agents phase II clinical trial program, including in patients with glioblastoma, NSCLC, melanoma, and small-cell lung, prostate, ovarian and breast cancers.	Galmarini, 2009
18	Thiflavin T		Amyloid imaging agent	Gan et al., 2013
19	Tiaramide		Anti-inflammatory agent	Tanaka et al., 2007.

Table 2. (Continued)

Comp No.	Name of the drug	Chemical structure	Biological action/use	Reference
20	Viozan (Sibenadet HCl)		Dual D2 dopamine receptor, β2-adrenoceptor agonist, in COPD	Ind et al., 2003
21	Zopolrestat		Aldose reductase inhibitor for treatment of diabetic complications, antidiabetic	Ding et al, 2009

Table 3. Chemical structure of some naturally occurring benzothiazoles isolated from terrestrial, microbial and marine sources

S. No.	Compound name and compound no	Biological source	Chemical structure	Reference
1	(8β,10α)-8-(angeloyloxy)-5,10-epoxythiazolo(5,4-α)bisabola-1,3,5,7-(14)-tetraene-4,11-diol (22)	Roots of the plant *Ligularia dentate*		Baba et al., 2007
2.	Benzothiazole (23)	Volatiles of American cranberries *Vaccinium macrocarpon* Ait. var. Early Black. Red deer *Cervus elaphus*, volatile fraction of French oak wood, tea leaves fungi *Aspergillus clavatus*		Anjou & Von Sydow, 1967; Vitzthum et al., 1975
3.	Cyclodercitin (24)	Deep-water marine sponge, *Dercitus* sp.		Gunawardana et al, 1988

Table 3. (Continued)

S. No.	Compound name and compound no	Biological source	Chemical structure	Reference
4.	Dercitin (25)	Deep-water marine sponge, *Dercitus* sp. Possess antitumour, antiviral, and immunomodulatory properties *in vitro*, and antitumour properties in vivo		Gunawardana et al., 1988
5.	Erythrazoles A (26)	Erythrobacter sp. from mangrove sediments		Hu & MacMilan, 2011
6.	Erythrazoles B (27)	Erythrobacter sp. from mangrove sediments		Hu & MacMilan, 2011

S. No.	Compound name and compound no	Biological source	Chemical structure	Reference
7.	6-Hydroxy benzothiazole-5-acetic acid or antibiotic C304A or M4582 (28)	Bacterium *Actinosynnema* sp and *Paecilomyces lilacinus*		Yaginuma et al., 1989
8.	Kuanoniamine A (29)	Mollusc *Chelynotus semperi*		Carroll & Scheuer, 1990
9.	Kuanoniamine B, C, D (30-32)	Mollusc *Chelynotus semperi*	Kuanoniamine B (R= $COCH_2CHMe_2$) Kuanoniamine C (R= $COCH_2Me$) Kuanoniamine D (R= COMe)	Carroll & Scheuer, 1990

Table 3. (Continued)

S. No.	Compound name and compound no	Biological source	Chemical structure	Reference
10.	D-Luciferin (33)	*Photinus pyralis*		White et al., 1963
11.	Mevashuntin (34)	*Streptomyces prunicolor*		Shin et al., 2005
12.	Oxyluciferin (35)	*Photinus pyralis*		White et al., 1963

S. No.	Compound name and compound no	Biological source	Chemical structure	Reference
13.	Rifamycins P(36) and Q (37)	*Nocardia mediterranei*	Rifamycin P (R=H) Rifamycin Q (R=CH$_2$OH)	Sensi et al., 1959
14.	Stellettamine (38)	Sponge *Stelletta* sp		Gunawardana et al., 1992
15.	S1319 (39)	Marine sponge *Dysidea* sp. Possess bronchodilatory activity		Suzuki et al., 1999

Table 3. (Continued)

S. No.	Compound name and compound no	Biological source	Chemical structure	Reference
16.	Thiazinotrienomycin F(40) and thiazinotrienomycin G (41)	*Streptomyces sp*	Thiazinotrienomycin F (R= cyclohex-1-enyl) Thiazinotrienomycin G (R= cyclohexyl)	Hosokawa et al., 2000
17.	Violatinctamine (42)	Genus *Cystodytes cf. violatinctus*		Chill et al., 2004

1.2. Naturally Occurring Benzothiazoles

This versatile biologically active scaffold has also shown its presence in various natural products isolated from marine and terrestrial plants (Table 3). The classical example of a naturally occurring benzothiazole is a light emitting pigment D-luciferin from the firefly. Seifert and King reported that benzothiazoles are also responsible for the aroma of tea leaves and cranberries (Seifert & King, 1982). Suzuki and coworkers in 1999 isolated a compound S1319 from the marine sponge possessing potent bronchodilatory activity (Suzuki et al., 1999.). Dercitin, a pyridoacridino alkaloid containing benzothiazole core, is a violet pigment isolated from a deep water marine sponge Dercitus *sp*. Dercitin and related compounds have been reported to possess antitumor, antiviral and immunomodulatory activities (Gunawardana et al., 1988). A group of highly colored pentacyclic alkaloids Kuanoniamine A-D were isolated in 1990 by Carroll and Scheuer from Micromesian purple tunicate and its predator. These compounds also display cytotoxic properties in addition to insecticidal and chelating properties (Carroll & Scheuer, 1990). Youcai and John in 2011 isolated the tetrasbustituted benzothiazoles bearing a diterpene side and glycine unit namely Erythrazoles A and B from erythrobacter *sp*., from mangrove sediments. The erythrazole B was found to be a potent cytotoxic agent against non-small cell lung cancer (NSCLC) cell lines (IC_{50}= 1.5-6.8 mM). (Hu & MacMillan, 2011)

1.3. Synthesis of Benzothiazole Derivatives

The first benzothiazole compound bearing substituent at 2nd position was synthesized by A. W. Hofmann in 1887 via simple cyclization mechanism. Since then, a number of synthetic routes have been developed and various methods for the convenient synthesis of benzothiazole derivatives have been reported in the literature. Among all the methods, the most efficient and one pot synthesis method to obtain benzothiazole derivatives in high yield involves condensation of 2-aminothiophenol with different aldehydes and carboxylic acid. (Scheme 1) (Victor et al., 2012; Ben-Alloum et al., 1997; Boger et al., 1978; Gill et al., 2015).

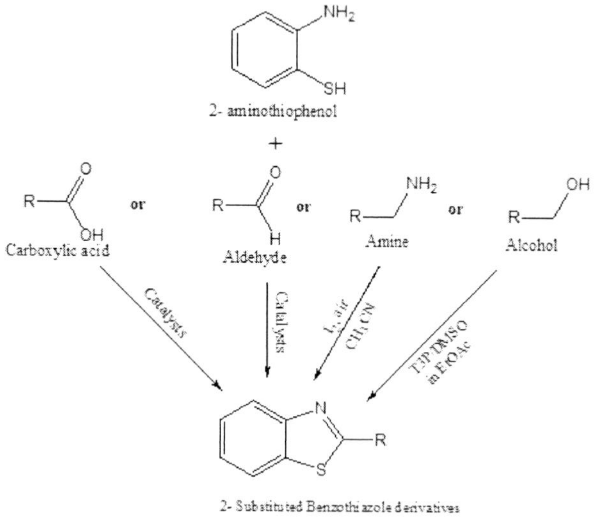

Catalysts with R-CHO: Iodine; PCC; Baker's yeast/CH$_2$Cl$_2$; Tungstophosphoric acid: Zirconium phospate (HTP/ZrP); RuCl$_3$; Cetyltrimethylammonium bromide (CTAB); H$_2$O$_2$ (30%)/HCl (37%) in EtOH; Na$_2$S$_2$O$_5$/DMF;

Catalysts with R-COOH: Phsophoric acid (PPA); Polyphosphate ester (PPE); P$_2$O$_5$/CH$_3$SO$_3$H

R= alkyl or aryl group

Scheme 1. One pot synthesis of 2- substituted benzothiazole derivatives using 2-aminothiophenol.

Benzothiazoles can also be synthesized in appreciable amounts by using various other alternative methods utilizing; (i) substituted thioureas (ii) thioformanilides (iii) substituted isothiocyanates and (iv) Weinreb amide reagent (Gill et al., 2015).

2. BIOLOGICAL ACTIVITIES AND USEFULNESS OF BENZOTHIAZOLES

Benzothiazole and its derivatives have been shown to exhibit wide spectrum of biological activites (Figure 2) which makes benzothizole nucleus a promising pharmacophore for drug development.

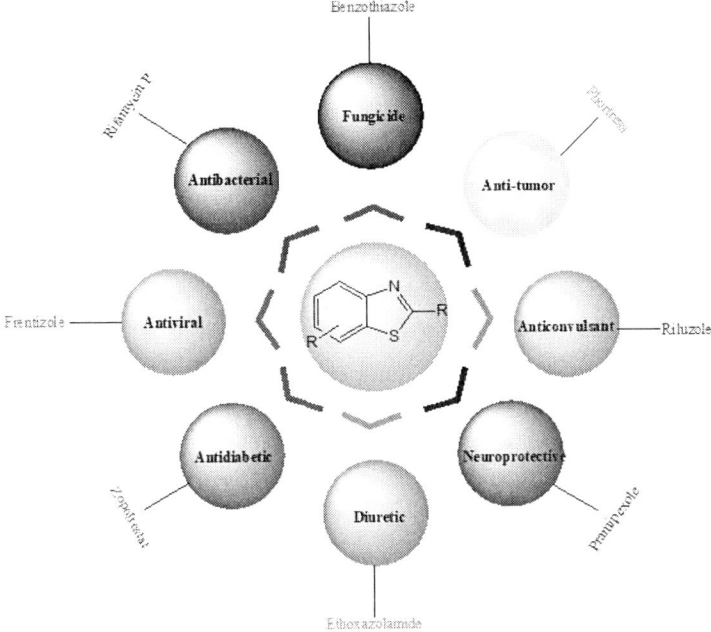

Figure 2. Biological activities spectrum of benzothiazoles.

Over the past few decades, medicinal chemists have synthesized and explored the therapeutic potential of several thousands of benzothiazole derivatives. The extensive investigations have led to the development of an anticonvulsant drug riluzole (Satyanarayana et al., 2008), anti-viral and immunosuppressant drug frentizole which is used clinically in rheumatoid arthritis (Scheetz et al., 1977), ethoxazolamide as diuretic (Supuran, 2008), zopolrestant, an aldose reductase inhibitor for the treatment of late-stage diabetic complications including neuropathy, nephropathy (Ding et al., 2009) and some anti-tumor drugs which are currently under different phases of clinical trials such as PMX-610, Phortress, DF 203 etc., (Bradshaw & Westwell, 2004) (Figure 2). Probenazole and methabenzthiazuron are used in agriculture as herbicide in winter corn crops while thioflavin T and some bis- benzothiazoles are excellent amyloid imaging agents (Reemtsma et al., 1995; Gan et al., 2013). The benzothiazole derivatives are also known to accelerate vulcanization and find an important application in rubber industry

(Fiehn et al., 1994). Benzothiazole salts are also used as sensitizing dyes in silver photography (Henary et al., 2013).

2.1. Benzothiazoles as Anti-Tumor Agents

In recent past, 2-substituted benzothiazoles have received considerable attention owing to their potent cytotoxic activity. The pharmacophore was rationally modified/substituted in an effort to overcome the limitations of the currently available chemotherapeutic agents and to develop the compounds with improved antitumor activity. The continuous efforts of scientific community working on this versatile pharmacophore culminated in development of benzothiazole based anticancer agents such as Phortress, PMX 610, etc., which have entered into various phases of clinical trials. These molecules are highly selective and potent class of antitumor agents active against lung, ovary, breast, renal and colon human cancer cell lines. Benzothiazole derivatives act on different molecular targets as outlined in the Figure3.

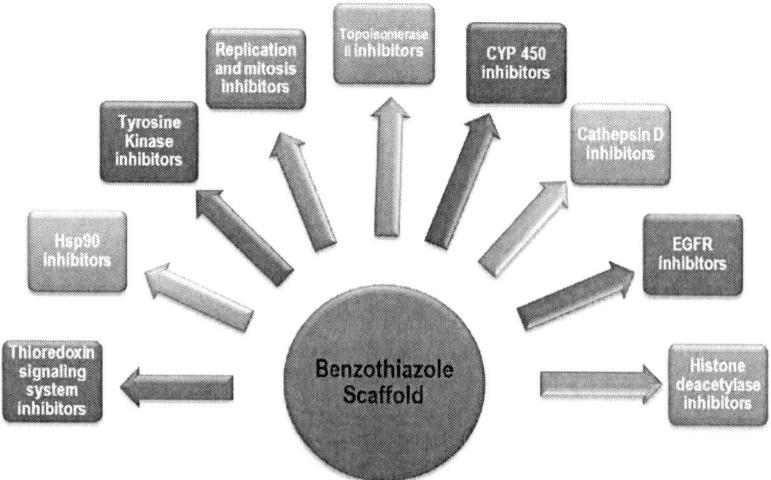

Figure 3. Molecular targets of benzothiazole derivatives as antitumor agents.

Table 4. List of patents on benzothiazole derivatives as anticancer agents

S. No	Patent no	Patent year	Inventors	Title
1.	US8143258 B2	2009	Okaniwa M et al.,	Benzothiazole compounds useful for Raf inhibition
2.	US20090118272 A1	2009	Chunjian Liu et al.,	Benzothiazole and azabenzothiazole compounds useful as kinase inhibitors
3.	WO100144	2010	Swinnen D et al.,	Fused bicyclic derivatives as PI3 kinase inhibitors and their preparation and use for the treatment of diseases
4.	WO007318	2010	Bacque E et al.,	Preparation of imidazo(1,2-a) pyrimidine derivatives as c-Met inhibitors and their pharmaceutical compositions
5.	WO089507	2010	Nemeceek C et al.,	Preparation of 6-(6-O-substituted triazolopyridazine sulfanyl) Benzothiazole and benzimidazole derivatives as c-Met inhibitors and their pharmaceutical compositions
6.	CN101759694	2010	Chen Z et al.,	Method for preparation of 2-(1,3-disubstituted phenyl-4-pyrazolyl)benzothiazoles and application as antitumor agent
7.	WO024903	2010	Feng Y et al.,	Preparation of benzo(d)oxazoles and benzo(d)thiazoles as kinase inhibitor
8.	WO064722	2010	Okaniwa M and Takagi T	Preparation of benzothiazole derivatives as anticancer agents.
9.	WO054058	2010	Apuy JL et al.,	Preparation of imidazobenzothiazoles as modulators of protein kinase.
10.	WO056764;	2011	Bhagwat SS	Preparation of imidazo (2,1-b)(1,3)benzothiazole derivatives for treatment of proliferative diseases.
11.	EP2358689 A1	2011	Okaniwa M et al	Benzothiazole derivatives as anticancer agents
12.	US7928140 B2	2011	Shon Booker	Benzothiazole PI3 kinase modulators for cancer treatement
13.	WO010715	2011	Kodama T et al.,	Pr-set7 inhibitor and therapeutic and/or preventive agent for cancer or life style-related disease.
14.	WO117882	2011	Kamal A et al.,	Preparation of pyrrolo(2,1-c)(1,4) benzodiazepine-benzothiazole or -benzoxazole conjugates linked through piperazine moiety for use in treating cancer.

Table 4. (Continued)

S. No	Patent no	Patent year	Inventors	Title
15.	INDE00270	2011	Kamal A et al	Benzothiazole hybrids useful as anticancer agents and process for the preparation thereof.
16.	WO121223	2011	Damour D.	Preparation of 6-((6-alkyl or 6-cycloalkyl-triazolopyridazin-3-yl) sulfanyl)-1,3-benzothiazole derivatives as c-Met inhibitors and their pharmaceutical compositions.
17.	INMU02510	2011	Malik JK et al.,	Novel diaryl substituted imidazo (2,1-b)benzothiazole derivatives, process for their preparation, and their pharmacological evaluation.
18.	WO153192	2011	Panicker B et al.,	Preparation of benzo(d)thiazole compounds as cytochrome P450 inhibitors and therapeutic uses thereof.
19.	WO110959	2012	Kamal A et al.,	Imidazolylphenyl) benzothiazole derivatives as anticancer agents and process for the preparation thereof.
20.	CN102391207	2012	Xue W et al.,	Preparation of N-(2-(substituted benzothiazole-2-carbamoyl)-phenyl)- benzamide derivatives as antiphytoviral and antitumor agents.
21.	WO104875	2012	Kamal A et al.,	Benzothiazole hybrids useful as anticancer agents and process for the preparation thereof.
22.	US0215154	2012	Wang JJ et al.,	Synthesis of 2-(4-aminophenyl) benzothiazole derivatives and use thereof.
23.	WO162463	2012	Jackson PF et al.,	Preparation of benzothiazolyl inhibitors of pro-matrix metalloproteinase activation.
24.	WO11016	2012	Kamal A et al.,	Benzothiazole derivatives as potential anticancer agents and apoptosis inducers and process for the preparation thereof.
25.	US8546393 B2	2013	Eva Albert et al	6-Triazolopyridazine sulfanyl benzothiazole derivatives as MET inhibitors
26.	WO055895	2013	Wang L et al.,	Preparation of substituted benzothiazoles as apoptosis-inducing agents for the treatment of cancer, immune and autoimmune diseases.
27.	US0178629	2013	Su TL et al.,	Synthesis of 4H-benzo(d)pyrrolo(1,2-a)thiazoles and indolizino(6,7-b)indole derivatives as antitumor therapeutic agents.
28.	CN103450176	2013	Li X et al.,	2-(4- Aminophenyl)benzothiazole-containing naphthalimide derivatives useful in treatment of cancers and their preparation.
29.	US8344007	2013	Tang JCO et al.,	Preparation of methylcantharimide derivatives as antitumor agents.

S. No	Patent no	Patent year	Inventors	Title
30.	CN103435574	2013	Li X et al.,	Preparation of mercapto benzothiazole derivatives as antitumor agents.
31.	KR128693	2013	Hong SH et al.,	Preparation of benzothiazole derivatives for the treatment of cancer.
32.	WO127345	2013	Zou Q et al.,	Preparation of heterocycles as protein kinase inhibitors.
33.	WO043002	2013	Hur Y et al.,	Imide-containing benzothiazole derivatives and pharmaceutical composition comprising the same.
34.	CN103772317	2014	Li Y et al.,	Predation of 2-methoxy-3-substituted-sulfonylamino-5-(2-acetamido-6-benzothiazole)-benzamide derivatives as antitumor agents.
35.	INDE00392	2014	Kamal A et al.,	2-Phenyl benzothlazole linked imidazole compounds as potential anticancer agents.

Many patents have been granted to benzothiazole derivatives exhibiting antitumor activity which further emphasize its significance as potential lead to develop antitumor agents. Few examples of the patent filed on benzothiazoles are presented in Table 4.

2.2. Development of Phortress- a Potent Antitumor Drug with Novel Mechanism of Action

Phortress (2-(4-amino-3-methylphenyl)-5-fluorobenzothiazole lysylamide dihydrochloride; NSC 710305) is a prodrug (Bradshaw & Westwell, 2004) that induces its own metabolism and which exhibits high selectivity against breast tumors, ovarian, lung, colon and renal carcinoma through novel mechanism of action. The drug was developed by a multidisciplinary research team led by Dr. Stevens at Nottingham University, UK through optimization of synthetic intermediate lead molecules.

- In 1970, CJM 126 (2-(4-aminophenyl) benzothiazole) in concentration less than 1 μM was observed to inhibit the growth of MCF-7 human-derived breast cancer cells. It GI_{50} was <10 nM.(Shi et al., 1996)
- To further improve the potency, a methyl group identified by structure activity relationship studies was introduced at the 3rd position in CJM 126 to obtain DF 203 ((2-(4-amino-3-methylphenyl) benzothiazole in 1995. Though, this molecule demonstrated superior *in vitro* and *in vivo* spectrum of antitumor activity but it suffered from metabolic instability leading to the formation of an inactive hydroxylated derivative. (Bradshaw et al., 2001).
- The problem of metabolic deactivation was resolved by the introduction of a fluorine atom in the benzene ring of benzothiazole i.e., by synthesizing the fluorinated analogue of DF 203. But the limited bioavailability owing to high lipophilicioty and poor

Benzothiazole 141

aqueous solubility of the optimized compound 5F 203 posed another challenge to the group (Hutchinson et al., 2001).

- To overcome the bioavailability issues, more water soluble derivatives of 5F 203 amide prodrug forms were synthesized by the conjugation with lysine and alanine amino acids. The superior pharmacokinetic profile, excellent antitumour activity (IC_{50} value for MCF-7 cells is 40 nM) and selectivity of lysyl amide prodrug (Phortress) made this the most suitable candidate for evaluation in clinical trials (Bradshaw et al., 2002).

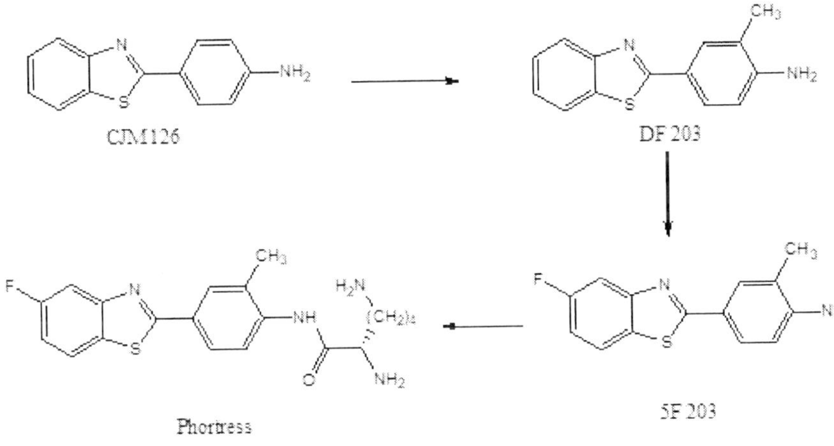

Figure 4. Development and optimization of Phortress.

2.2.1. *Mechanism of action of Phortress*

Phortress undergoes hydrolysis to release 5F 203 which then binds to the aryl hydrocarbon receptor (AhR) within sensitive cells to cause induction of CYP1A1 enzyme. The enzyme metabolizes 5F 203 to generate highly reactive elecrophilic species at the tumor site which covalently bind to the cell's DNA resulting in single- and double-DNA strand breaks and ultimately cell death (Figure 5). (Chua et al., 2000; Leong et al., 2003; Leong et al., 2004). The systemic toxicity of Phortress is minimal in comparison to other clinically available chemotherapeutic agents as it targets mainly cancer cells wherein it causes induction and expression of CYP isoforms.

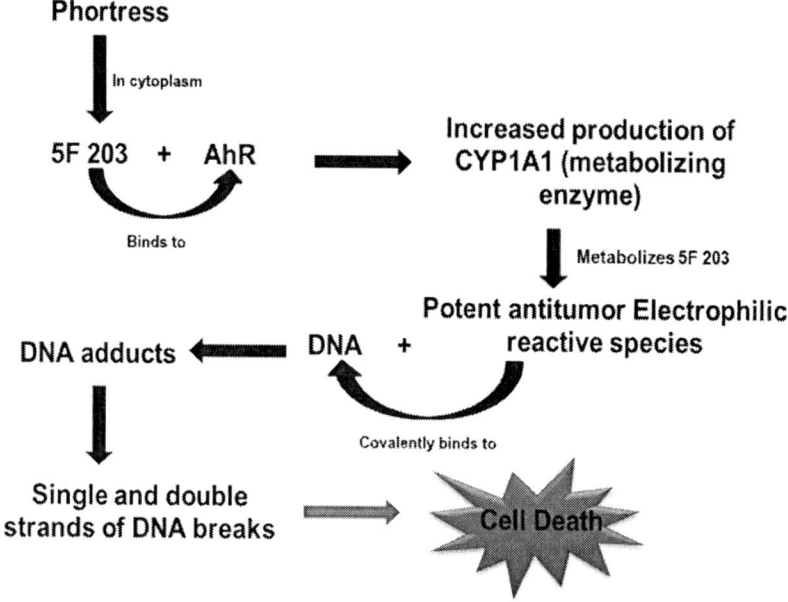

Figure 5. Mechanism of action of Phortress.

2.3. Recent Advances in the Development of Antitumor Benzothiazole Derivatives

Since the development of Phortress and other potent chemotherapeutic agents bearing a benzothiazole nucleus, numerous new benzothiazole derivatives have been synthesized through the structural modification but primarily by varying the functional groups at the C-2 and C-6 positions of the pharmacophore. These benzothiazoles derivatives have been shown to exert their anticancer activities by acting on various biological targets. Some important examples of recently synthesized benzothiazoles displaying potential cytotoxic activity by interacting with the biomolecular targets are discussed below.

2.3.1. Protein Kinase Inhibitors

2.3.1.1. B- Raf Inhibitors

B-Raf is a kinase protein more commonly known as serine/threonine-protein kinase B-Raf. It is involved in growth signal transduction inside cells. It has been shown to be mutated in certain type of cancers. The Raf mutants cause cancer by excessively signally cells to grow. The FDA approved B-Raf inhibitors like sorafenib and vemurafenib for the treatment of cancer initiated great interest in developing chemotherapeutic agents which target the mutant B-Raf (Tsai et al., 2008). Okinawa and Takagi in 2010, reported the synthesis of some amide derivatives of benzothiazole as anticancer agents. The most potent compound **43** at 1.0 µM showed strong inhibitory rate of 101% against Raf Kinase. The compound could be developed as a cancer growth inhibitor and a cancer metastasis suppressive agent (Okaniwa & Takagi, 2010).

El Damasy and co-workers in 2016 designed and synthesized structural analogs of anticancer drug, sorafenib. In the structurally modified benzothiazole amide and urea derivatives, the pyridylamide moiety was linked through ether linkage at the 6th position of benzothiazole ring. The central phenyl ring of sorafenib was replaced with benzothiazole scaffold while retaining the other main structural features with an aim to improve the cellular anticancer activity by selective inhibition of B-Raf^{V600E} and C-Raf kinases. Compound **44** showed remarkable superior potency and efficacy compared to sorafenib as well as notable extended spectrum activity covering 57 human cancer cell lines. Further, Compound **44** at concentration of 10 µM in kinase screening over 10 oncogenic kinases showed selective inhibitory activities towards B-Raf^{V600E} and C-Raf kinases. However, compound **45** showed better potencies than **44** as RAF inhibitor, with nanomolar IC$_{50}$ values. These results of enzymatic assay were well supported by the molecular docking study. Compounds, **44** and **45** were also found to have the low affinities towards the different CYP450 isoforms and hERG channel, and thus low possibility for drug-drug interactions and cardiac side effects. SAR studies indicated the both urea spacer and disubstituted phenyl group are essential for exhibiting the best anticancer activity (El-Damasy et al., 2016). They further modified the structure of a potent anticancer uriedobenothiazole (compound **46**) by incorporating a (4-ethyl piperazin-1-yl) methyl as a hydrophobic tail to improve its physicochemical properties and thus anticancer potency. It led to the discovery of Raf kinase inhibitor (compound **47**) (KST016366; 4-((2-(3-(4-((4-ethylpiperazin-1-yl) methyl)-3-(trifluoromethyl) phenyl) ureido) benzo(d)thiazol-6-yl) oxy) picolinamide) as a new potent multikinase inhibitor. This compound displayed substantial broad-spectrum antiproliferative activity against 60 human cancer cell lines particularly against leukemia K-562 and colon carcinoma KM12 cell lines (GI_{50} values of 51.4 and 19 nm, respectively). It was having good oral bioavailability and also found to selectively inhibit certain oncogenic kinases implicated in both tumorigenesis and angiogenesis in nanomolar concentration. This molecule may serve as a lead for further development of potent anticancer agents (El-Damasy et al., 2016).

2.3.1.2. Tyrosine Kinase Inhibitors (TKI)

Tyrosine kinases are enzymes responsible for the activation of many proteins by signal transduction pathways. The TKIs inhibit the phosphorylation of proteins and thus block their activation. In recent years, tyrosine kinase has emerged as an effective target for the treatment of various cancers. Various TKIs inhibitor drugs developed to treat malignancies include imatinib, gefitinib, erlotinib, sorafenib, sunitinib, and dasatinib etc. Bhuva and Kini in 2010 reported the synthesis of a series of novel 2-phenyl-1,3-benzothiazoles (compounds **48a-l;** Figure 6). The prepared compounds upon testing for their anticancer activity against MCF-7 breast cancer cell line with the MTT assay were found to mimic the ATP competitive binding of genistein and quercetin to tyrosine kinase. The compounds exhibited moderate to good antibreast cancer activity (Bhuva and Kini, 2010).

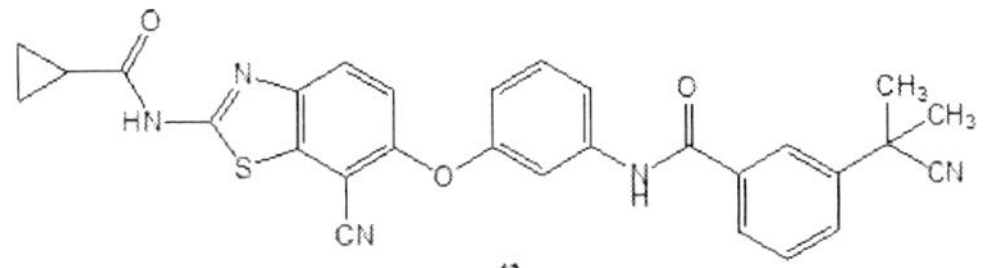

43

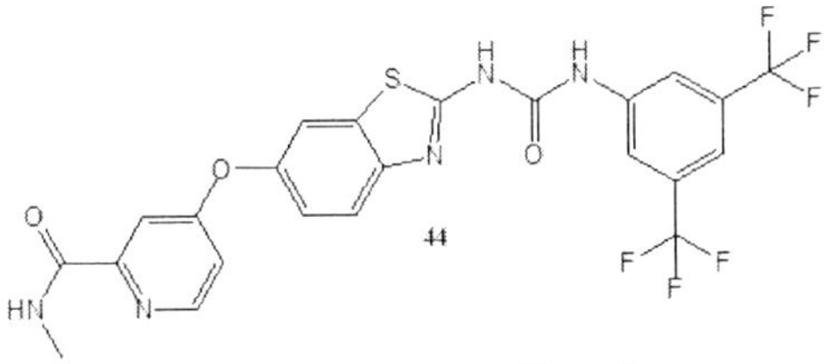

44

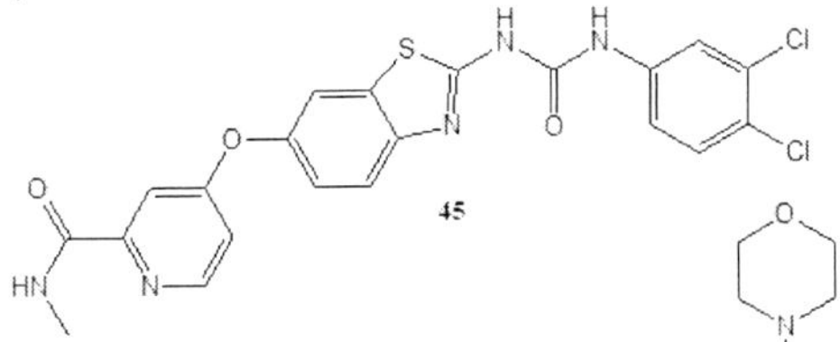

45

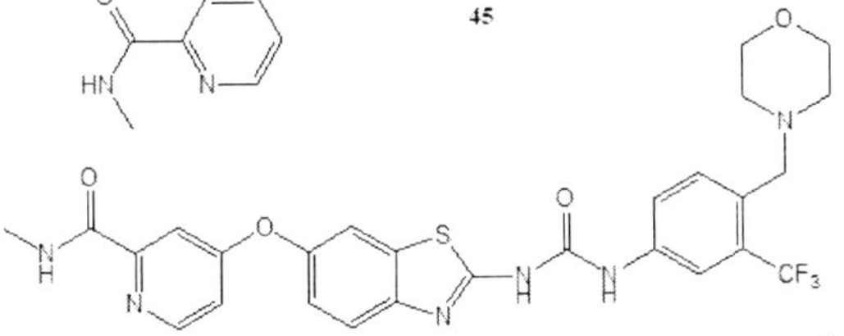

46

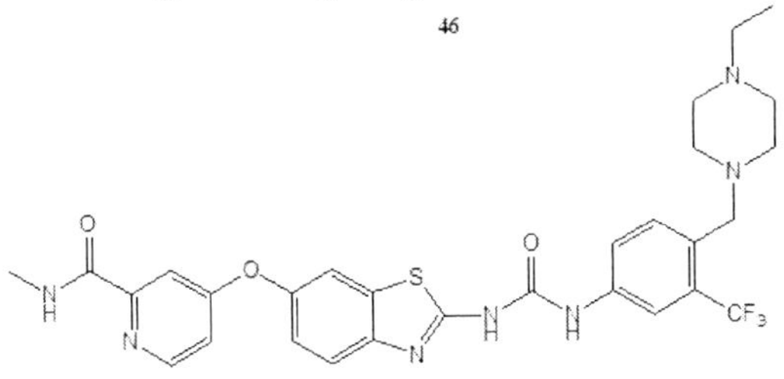

47

Figure 6. Chemical structure of benzothiazole derivatives as Raf inhibitors.

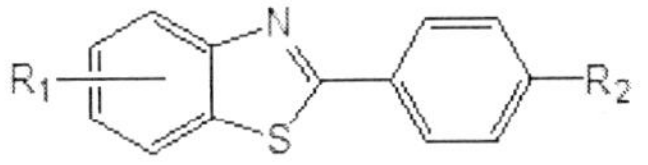

48a:R_1=6-F;	$R_2=CH_3$	**48g**:R_1=7-Br;	$R_2=OCH_3$
48b:R_1=5,6 diCH$_3$;	$R_2=CH_3$	**48h**:R_1=5Cl;	$R_2=OCH_3$
48c:R_1=6-Br;	$R_2=OCH_3$	**48i**:R_1=7-Cl;	$R_2=OCH_3$
48d:R_1=6-F;	$R_2=OCH_3$	**48j**:R_1=4-F;	$R_2=OCH_3$
48e:R_1=6-Br;	$R_2=CH_3$	**48k**:R_1=5,6-di F;	$R_2=OCH_3$
48f:R_1=5-Br;	$R_2=OCH_3$	**48l**:R_1=4-F;	$R_2=CH_3$

Figure 7. Chemical structure of benzothiazole derivatives as Tyrosine Kinase inhibitors.

2.3.1.3. Vascular Endothelial Growth Factor Receptor (VEGFR) Kinase Inhibitors

Angiogenesis is primarily controlled by various signal transduction pathways. Vascular endothelial growth factor (VEGF)/vascular endothelial growth factor receptor (VEGFR) plays an important role in angiogenesis. Tumor cells secrete VEGF which can support new blood vessels growth to supply oxygen and nutrients and maintenance of a vascular network. This results in promotion of metastatic spread of primary tumor cells to distant sites in the body. Several studies have proved that VEGFR-2 is the major receptor for angiogenesis (Alitalo & Carmeliet, 2002). Inhibition of VEGF/VEGFR system leads to interference in signal transduction pathways and seems to be a promising approach to target the tumor cells and to impede the tumor growth and metastasis (Ferrara, 2005).

Tasler et al., (Tasler et al., 2009) reported the synthesis and *in vitro* inhibitory activity on 16 kinases of various *N*-substituted 21- (3-aminophenyl)- and 21-(4-aminophenyl)-benzothiazoles. 4'-(aminoaryl benzothiazole)- 6',7'-dimethoxyquinazoline (**49a**) because of its potent activity was used for further optimizations. They prepared di-(hetero) arylether series to identify a lead molecule (Comp **49b**) based on the hit structure for kinase inhibition, (Tasler S et al., 2009). This compound was found to have enzymatic IC_{50} values < 100 nM, cellular EC_{50} values between 60 - 800 nM and therefore possessed excellent EC_{50}/IC_{50} ratios on selected kinases, Interestingly, ether linkage compounds (**49c**) inhibited VEGFR kinases as well as tyrosine kinase with immunoglobulin and EGF homology domains 2 (TIE2). On the other hand, the corresponding NH-derivatives (**49 a,b**) displayed excellent inhibitory potential on EGFR/ErbB2 (IC_{50} = <10 nM).

Kamal et al., in 2013 designed a new series of benzothiazole-pyrrole-based conjugates and evaluated them for cytotoxic activity against MCF-7 cell line (Kamal, et al., 2013). Two compounds **50a** and **50b** amongst all were the most effective in inducing apoptosis in MCF-7 cells. Compound **50a** also showed down regulation of oncogenic expression of Ras and its downstream effector molecules such as MEK1, ERK1/2, p38MAPK, and VEGF. A new series of benzothiazole Schiff bases prepared by Gabr et al.,

(Gabr et al., 2016.) were tested for their antitumor activity against cervical cancer (HeLa) and kidney fibroblast cancer (COS-7) cell lines. Compounds **51 a–e** showed promising activity against HeLa cell line with IC_{50} values of 2.41, 3.06, 6.46, 2.22, and 6.25 µmol/L, respectively, in comparison to doxorubicin (IC_{50}= 2.05 µmol/L). The most potent compound **51a** showed comparable activity (IC_{50} =4.31 µmol/L) with respect to doxorubicin against COS-7 cell line (IC_{50} = 3.04mol/L). SAR study indicated that the presence of 2-(4-hydroxy-2-methoxybenzylidene) hydrazino moiety at the 2-position of benzothiazole nucleus is responsible for the enhanced activity against cervical cancer (Hela) and kidney fibroblast cancer (COS-7) cell lines. Both the compounds **51a** and **51b** were found to form two hydrogen bonds with the EGFR tyrosine kinase active site in docking studies.

Zhu and coworkers in 2017 designed and synthesized 6-chloro-*N*-(3,4-disubstituted-1,3-thiazol-2(*3H*)-ylidene)-1,3-benzothiazol-2-amine derivatives to target colorectal cancer cells. The rational for the synthesis was derived from the chemical structures of Erlotinib, Lapatinib, and Gefinitib. The most potent compound was found to be (6-chloro-*N*-(4-(3-chlorophenyl)-3-(4-nitrophenyl)-2,3-dihydro-1,3-thiazol-2-yl)-1,3-benzothiazol-2-amine) **(52)** which showed potent activity against all the colon cell lines. Molecular docking studies revealed that a hydrophobic group substitution is the primary structural requirement for fitting into the active pocket of the receptor, with 2 hydrogen bond acceptors and 1 hydrogen bond donor. They concluded that substitution of thiazole at 3rd position with the electron withdrawing group or introduction of hydroxyl group in the molecule lead to an increase in anticancer activity (Zhu et al., 2013). Zhang et al., in 2017 reported the efficient synthesis of some novel indole derivatives as selective and potent inhibitors for the VEGFR-2 tyrosine kinase. They found that presence of benzothiazole motif is associated with increased activity towards VEGFR-2. The most potent compound **53**; (*N*-(6-Methoxybenzothiazol-2-yl)-1*H*-indole-3-carboxamide) at 10 µM concentration demonstrated the highest growth inhibition rate of 66.7% against the VEGFR-2 tyrosine kinase. The molecular docking results showed that the VEGFR-2 tyrosine kinase-compound **53** complex was stabilized by the hydrogen bond between Cys

1045 amino acid residues of the protein and compound **53** (Zhang et al., 2017).

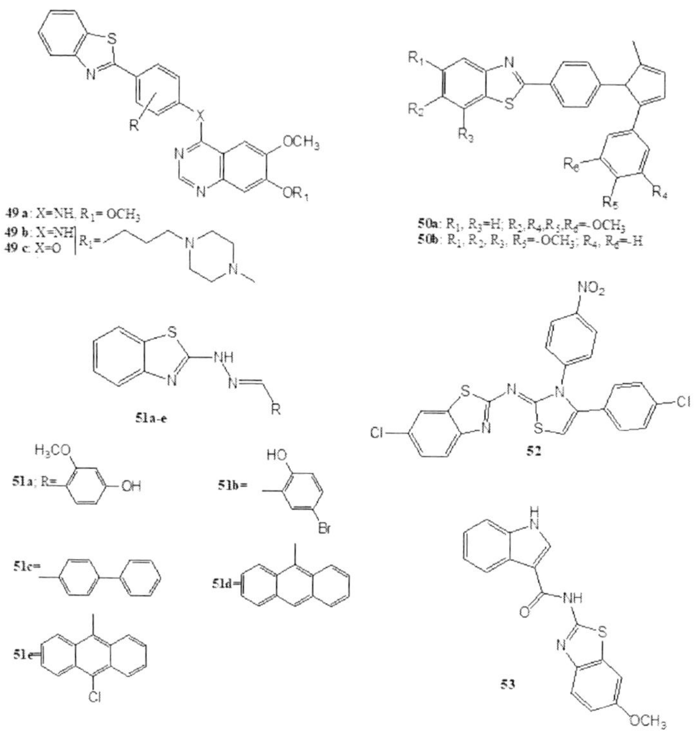

Figure 8a. Chemical structure of benzothiazole derivatives as VEGFR inhibitors.

Reddy et al., reported the synthesis and cytotoxic activity of a series of new pyrazolo-benzothiazole hybrids. The compounds were tested against cancer cell lines (colon (HT-29), prostate (PC-3), lung (A549), glioblastoma (U87MG)) and normal human embryonic kidney cell line (Hek-293T). Although several compounds exhibited significant activity but the compound **54** showed better activity than the axitinib. Compound **54**(*N*-(6-chlorobenzo(d)thiazol-2-yl)-3-(4-fluorophenyl)-1-phenyl-1*H*-pyrazole-4-carboxamide) was active against all the tested cancer cell lines with IC$_{50}$ values in the range 3.17-6.77 µM. Compound **54** also showed the strongest growth inhibition in 3D multicellular spheroids of PC-3 and U87MG cells. It displayed significant in vitro (VEGFR-2 inhibition) and *in* *vivo* (transgenic zebrafish Tg (flila:EGFP) model) antiangiogenic properties. The compound **54** was found to fit well in the AXI binding pocket of VEGFR-2 in the molecular docking studies. The carbonyl group showed hydrogen bonding with the peptide and side chain amide N-H of Asn923. The side chain N-H also interacts with the nitrogen atom present in the benzothiazole ring. The amide N-H present in the compounds interacts with the carbonyl oxygen of Leu840. The molecule also displays strong hydrophobic interactions with VEGFR-2. The substituted benzo(*d*)thiazol-2-amine scaffold show π-interactions with Phe1047 and hydrophobic interactions with the Gly841, Arg1032 and Asp1056 residues. This benzothiazole derivative could be used for further development and discovery of anticancer and antiangiogenic drug molecules (Reddy et al., 2019).

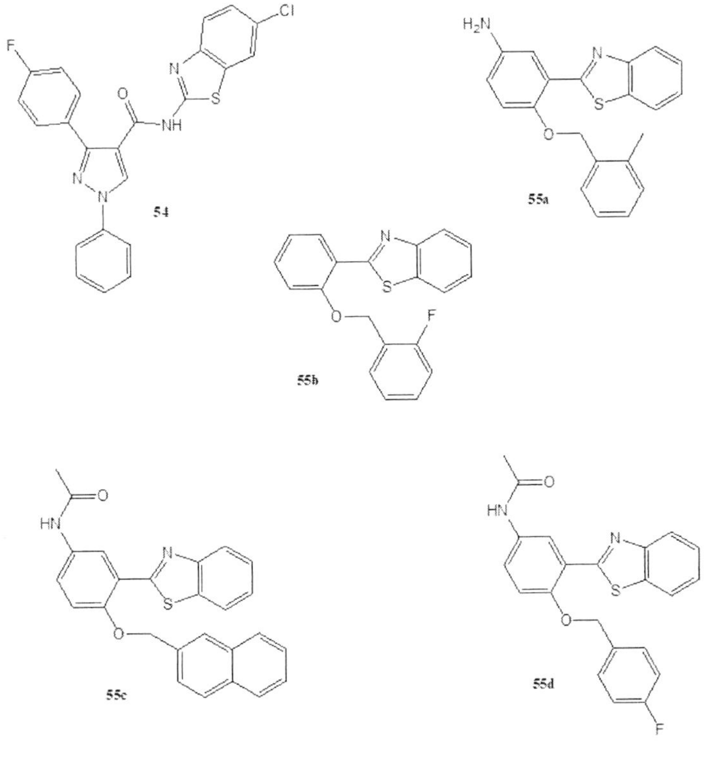

Figure 8b. Chemical structure of benzothiazole derivatives as VEGFR inhibitors.

A similarity/substructure-based search of eMolecule database was performed by Shahare and Talele in 2019 to discover promising benzothiazole derivatives as EGFR tyrosine kinase inhibitors. Molecular docking, pharmacokinetics and synthetic accessibility criteria were used on approximately 7000 molecules containing benzothiazole scaffold. A total of four molecules (**55a-d**) were identified to be promising EGFR tyrosine kinase inhibitors. Molecular docking indicated that the EGFR protein becomes stable upon binding of benzothiazole derivatives to the receptor cavity. Results of strong binding affinity of all molecules toward the EGFR target were supported by the binding energy calculations (Sahare and Talele, 2019).

2.3.1.4. c-Met Inhibitors

c-Met is a membrane receptor which is also known as tyrosine-protein kinase Met or hepatocyte growth factor receptor (HGFR). It plays crucial role in wound healing and embryonic development but its aberrant activity is associated with angiogenesis and thus helps in supplying nutrients to the tumor and metastasis of cancer cells (Gentile et al., 2008). It is an effective target in anticancer therapies during tumor evolution and resistance to treatment.

A novel compound **56** in which benzothiazole scaffold is attached to imidazopyrimidine moiety via a bridge sulfur atom was synthesized by Bacque et al., in 2010 as c-Met inhibitor (IC_{50} = <100 nM) (Bacque et al., 2010). Based on the similar strategy, imidazopyrimidine moiety was replaced with triazolopyridazinyl group to obtain potent c-Met inhibitors. Compound **57** exhibited good inhibition of c-Met with an IC_{50} value lower than 100 nM (Nemececk et al., 2010; Damour, 2011). Few new organic amino acid amides containing 4-(imidazo(*2,1-b*)benzothiazol-2-yl)phenyl moiety were designed by Furlan and coworkers as inhibitors of Met receptor tyrosine kinases. These compounds (**58**) were found to block Met-triggered cell effectively and noted to be quite effective agents targeting Met-driven tumorigenesis (Furlan, et al., 2012).

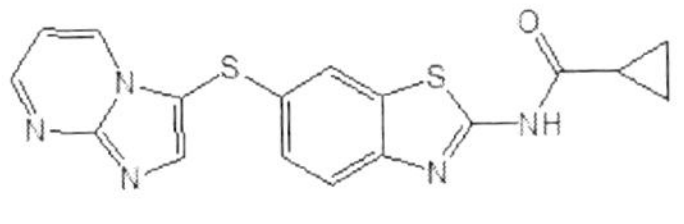

Benzothiazole 151

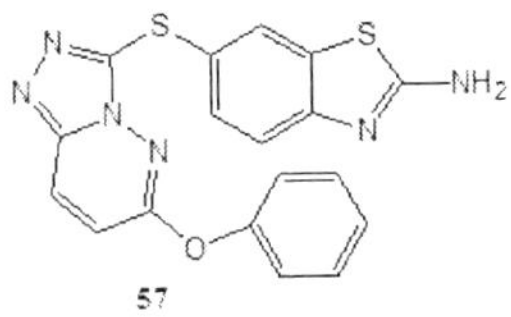

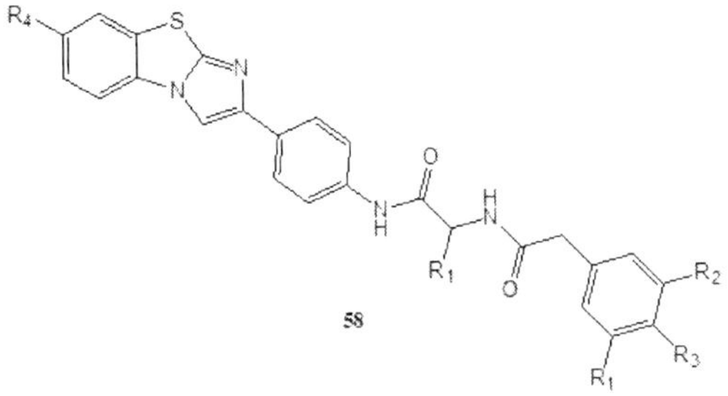

Figure 9. Chemical structure of benzothiazole derivatives as c-Met inhibitors.

2.3.1.5. Phosphoinositide-3-kinase Inhibitors

The uncontrolled cell growth in cancer is associated with dysregulation of the delicate network of cell signaling pathway. Phosphoinositide-3-kinase (PI3K)/AKT/mTOR pathway is one of the important signaling pathways that regulate important cellular functions such as growth control, metabolism and translation initiation. Its dysregulation lead to uncontrolled cell growth. PI3K inhibitors are known for tumor suppression and therefore it is considered as a potential target for the development of anticancer agents (Kurtz & Ray-Coquard, 2012). Few fused bicyclic derivatives as inhibitors of PI3K were patented by Swinnen et al., in 2010. The most potent compound, 2-acetamido benzothiazole **59** with sulfonamidoaryl substitution at 6th position inhibited PI3K significantly (IC_{50} = < 1 and < 5 µM (Swinnen et al., 2010).

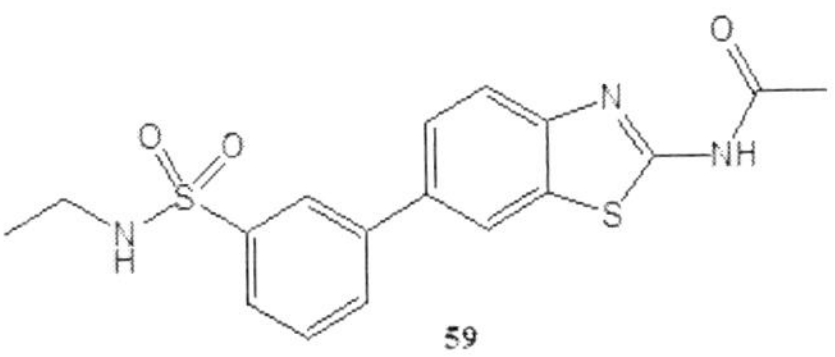

Figure 10. Chemical structure of a benzothiazole derivatives as Phosphonositide 3-kinase inhibitor.

2.3.2. Cytochrome p450 (CYP) Inhibitors

The cytochrome P450 (CYP) is a group of enzymes. These heme-thiolate enzymes are involved in the metabolism of a numerous xenobiotics and endogenous compounds such as hormones. Their ability to metabolize carcinogens, pro-carcinogens, and chemotherapeutics gives them an indisputable role in the cancer prevention and treatment strategies. Various studies have shown them to play a role in tumor formation and development. CYPs 1B1 and 2W1 are indeed expressed specifically in tumors. Therefore, inhibition of CYPs is a widely explored area of research for the treatment and prevention of cancer. (Gibson et al., 2003; Murray et al., 1997; Murray, 2000; Bruno et al., 2007).

Considering the crucial role of cytochrome P450 in cancer, some new fluorinated 2-(4-aminophenyl) benzothiazoles were demonstrated to be potently cytotoxic (GI_{50}< 1 nM) in sensitive human breast MCF-7 (ERþ) and MDA 468 (ER-) cell lines by Hutchinson et al., (Hutchinson et al., 2001). Most potent derivative was 2-(4-amino-3-methylphenyl)-5-fluoro-benzothiazole **5F 203**. Earlier they synthesized a series of water-soluble L-lysine- and L-alanyl-amide prodrugs of the lipophilic antitumour 2-(4-aminophenyl) benzothiazoles (**5F 203**). The prodrugs exhibited the required pharmaceutical properties of good water solubility (in weak acid) and stability at ambient temperature and degradation to free base *in vivo*. The lysyl-amide of 2-(4-amino-3-methylphenyl)-5-fluorobenzothiazole (Phortress) **11** is currently undergoing clinical trials (Hutchinson et al., 2002). Later in 2003, they reported the synthesis of a new series of antitumour 2-(4-aminophenyl) benzothiazole analogues substituted with cyano and alkynyl group at 3′ position. Compound **60** was active against MCF-7 and MDA-468 with GI_{50} value 0.0057 and 0.0078 µM, respectively, and compound **61** depicted GI_{50} values 0.011 and 0.0088 µM against MCF-7 and MDA-468, respectively (Hutchinson et al., 2003).

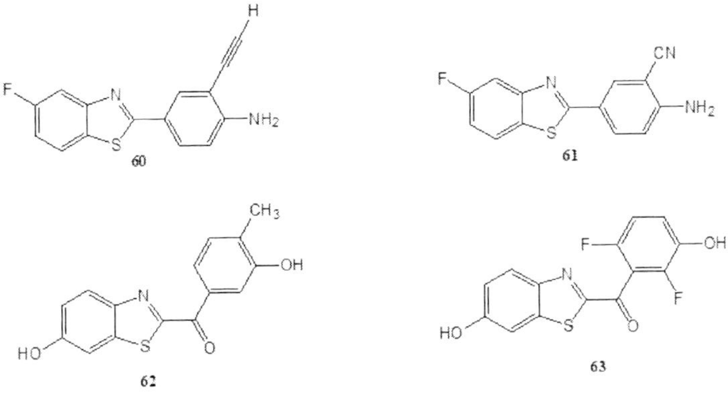

Figure 11. Chemical structure of benzothiazole derivatives as Cytochrome P450 inhibitors.

Spadaro et al., synthesized a library of benzothiazoles for inhibition of human 17β-HSD1, 17β-HSD2 and human hepatic CYP enzymes. Compounds (**62, 63**) exhibited high intracellular activity and good potency toward the target. Compound (**62**) inhibited CYP enzymes with an IC_{50} ¼ 0.57 mM. Compounds with fluorine-substitution increased, inhibitory activity in 17β-HSD1, 17β-HSD2 and introduction of the methyl group in the 4-position of **62** not only increased activity but also decreased inhibition of 17β-HSD2 (Spadaro et al., 2012).

2.3.3. *Histone Deacetylase (HDAC) Inhibitors*

The histone deacetylase inhibitors are a new class of cytostatic agents that inhibit the proliferation of tumor cells in culture and *in vivo* by inducing cell cycle arrest, differentiation and/or apoptosis. They exert their anti-tumour effects through modulating the acetylation/deactylation of histones and/or non-histone proteins such as transcription factors to cause induction of expression changes of oncogenes.

Han and coworkers obtained the patent for benzothiazole-based hydroxamic acid derivatives as inhibitors of histone deacetylase. The compound **64** showed excellent cytotoxicity (IC_{50} = 0.81 μg/ml) against few human cancer cell lines (Han et al., 2012).

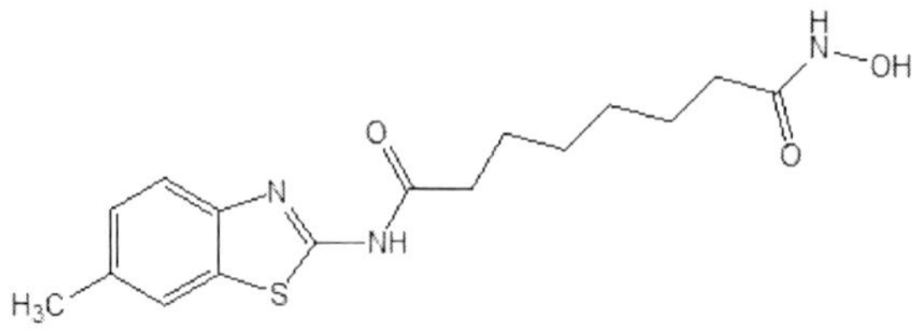

Figure 12. Chemical structure of a benzothiazole derivative as HDAC inhibitor.

2.3.4. Topoisomerase Inhibitors

DNA topoisomerase I and II inhibitors are commonly used antineoplastic agents in chemotherapy. Topoisomerases (TOP) are unique enzymes which are involved in various processes of DNA metabolism. These are necessary for the survival of all unicellular and multicellular organisms. This family of enzymes has generated considerable interest since many antimicrobial and anticancer drugs target topoisomerases and inhibit key steps in their catalytic cycle. Benzothiazole derivatives owing to their similarity with the structure of DNA purine bases, chemists have investigated the potential of their topoisomerase inhibition activities.

Pinar and his team reported the synthesis of substituted benzothiazole derivatives and evaluated their eukaryotic DNA topoisomerase II inhibitory activity in cell free system. 2-Phenoxymethyl benzothiazole (**65**) was observed to be more potent than the standard drug etoposide (Pinar et al., 2004). The topoisomerase I & II inhibitory activities of hydroxybenzoyl-2-amino benzothiazoles were evaluated by Abdel Aziz et al. in 2004. Three compounds *viz*. trihydroxybenzoyl-2-aminobenzothiazole (**66**) and dihydroxy derivative (**67**) demonstrated an IC_{50} value of 8.2 and 16.9 mM, but monohydroxy derivatives (**68**) had no topo I inhibitory activities. SAR studies suggested that inhibitory activity increases with an increase in number of hydroxy groups on phenyl ring (Abdel-Aziz et al., 2004).

Choi et al., used a combinatorial method to synthesize 2-(substituted-phenyl) benzothiazoles as antitumour agents capable of inhibiting topoisomerase II. Based on the structure activity relationship studies, they reported that benzothiazole scaffold is must for exhibiting potent cytotoxicity. They noted that substitution at 3-position of the phenyl ring with alkyl or halogen groups led to an increase in the cytotoxicity of synthesized compounds. The two most potent benzothiazole analogs **69** (2-(3-amino-4-methylphenyl)benzothiazole) and **70** (2-(3,4 dichlorophenyl)benzothiazole) were observed to possess the strongest inhibitory activity against topoisomerase II with the IC_{50} of 71.7 and 70.5 μM, respectively with respect to standard antitumor agent etoposide (IC_{50} of 78.4 μM.). Compounds **69** and **71** possessing amino substitution also showed high topo all activity (Choi et al., 2006).

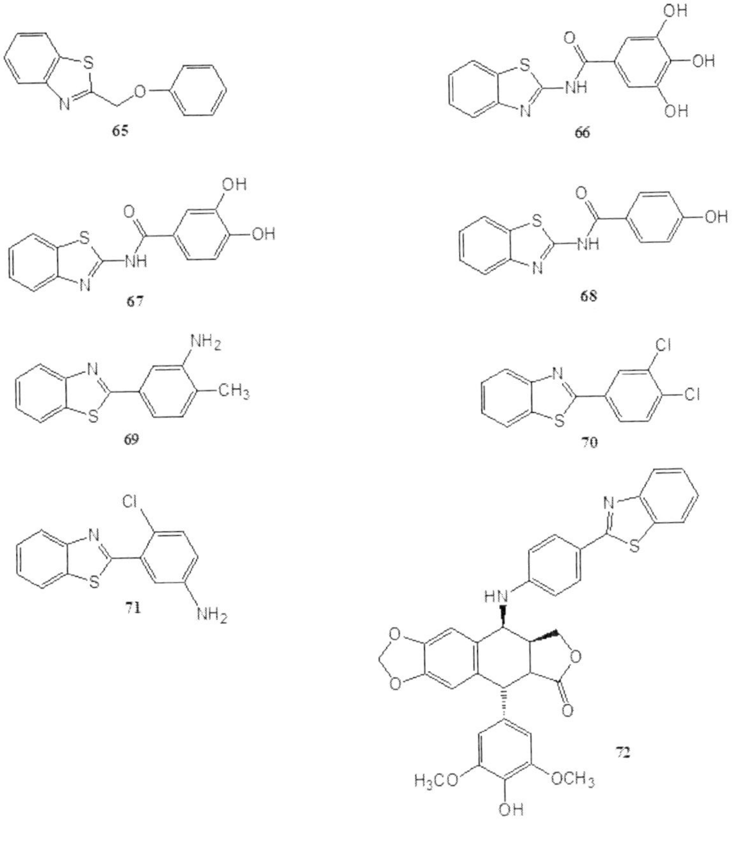

Figure 13a. Chemical structure of benzothiazole derivatives as topoi somerase inhibitors.

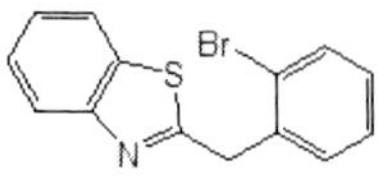

73

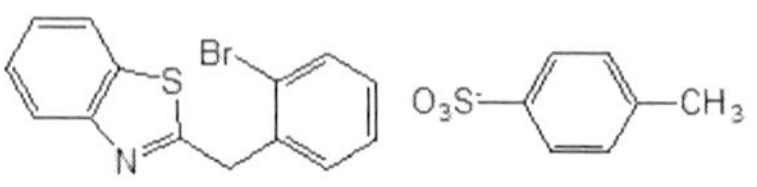

74

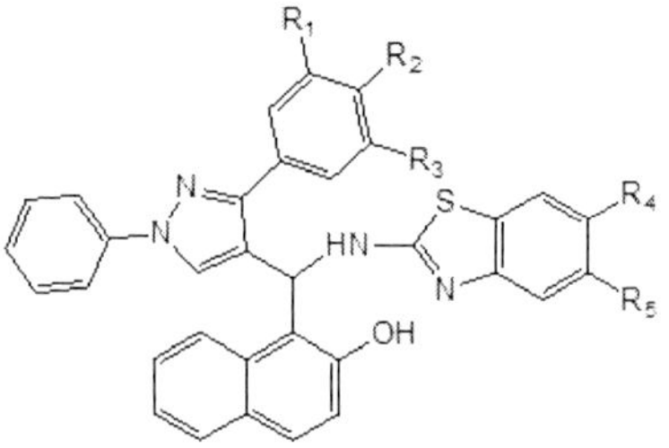

75a: R_1, R_3, R_5=H, R_4=OCH_3; R_2=Cl
75b: R_1, R_3, R_5=H, R_4=OCH_3; R=NO_2
75c: R_1, R_3, R_5=H, R_4=OCH_3; R_2=F

Figure 13b. Chemical structure of benzothiazole derivatives as topoi somerase inhibitors.

Podophyllotoxin belongs to lignane class of natural products. Sincere efforts were made to develop this as antitumour agent but because of its severe gastrointestinal side effects it failed in clinical trials. Kamal et al., in 2011 prepared the podophyllotoxin-benzothiazole conjugates by linking through an aryl amino spacer at C-4 position of the podophyllotoxin scaffold. The conjugates were investigated for their cytotoxicity on selected human cancer cell lines (Zr-75-1, MCF-7, KB, Gurav, DWD, A2780, Colon, A549, PC3, HT1080, DWD, Hop-62 and SiHa), and they exhibited significant cytotoxic activity against colon (Colo 205), lung (Hop-62) and oral (HT1080, DWD) cancer cell lines. Amongst all, one of the conjugates 4'-O-demethyl-4b- (4-(1,3-benzothiazole-2-yl)anilino)-4-desoxypodophyllotoxin derivative **72** was found to be more potent (IC_{50} values of 2.7 (colon 205) and 2.3 μM (DWD) than adriamycin and podophyllotoxin (IC_{50}; 5.1 and 5.0 μM, respectively). Compound **72** significantly inhibited the DNA topoisomerase-II activity suggesting that the bulky substitution at C-4 position is capable of potentiating the anticancer activity of such compounds (Kamal et al., 2011).

Ozen et al., synthesized and screened 2-(substituted phenyl/benzyl) benzothiazoles and their salts for anticipated eukaryotic DNA topoisomerase

I and II inhibitory activity. The *N*-amino tosylated salt form (**73**) of 2-bromobenzyl benzothiazole (**74**) was found to be the most effective derivative with an IC$_{50}$ value of 39.4 nM and was significantly more potent than etoposide (Ozen et al., 2013).

Nagaraju et al., designed and synthesized a library of thirty new pyrazole linked benzothiazole-β-naphthol derivatives using a simple, efficient and eco-friendly method under catalyst-free conditions. Three compounds **75a**, **75b** and **75c** produced significant cytotoxicity with IC$_{50}$ values ranging between 4.63 and 5.54 µM against human cervical cancer cells (HeLa). These derivatives were found to induce cell cycle arrest in G$_2$/M phase, possess good DNA binding affinity and can inhibit the topoisomerase I activity. Viscosity studies and molecular docking studies demonstrated that the derivatives bind to the minor groove of the DNA. The hydroxyl groups on naphthyl ring and benzothiazole amino functional groups were shown to form the hydrogen bond with the target enzyme. SAR studies indicated that substitution of benzothiazole at 6th position with electron activating group such as methoxy or methyl substituents increases the cytotoxicity while presence of electronegative groups such as F, Cl and NO$_2$ substituents on 4th position of phenyl ring attached to pyrazole was crucial for activity (Nagaraju et al., 2019).

2.3.5. Replication and Mitosis Inhibitors

Mitotic inhibitors bind to tubulin, a microtubular protein involved in cellular processes which play an essential role in mitosis, especially in induction of apoptosis. The inhibition of polymerization and assembly of microtubules result in disruption of mitotic spindles, cell cycle arrest and cell death (Kavallaris, 2010). Therefore, the discovery of potential microtubule inhibitors appears to be a fruitful strategy for the development of chemotherapeutic agents.

Tuylu et al., prepared and evaluated the mutagenicity of 2-aryl-substituted (*o*-hydroxy-, *m*-bromo-, *o*-methoxy-, *o*-nitro-phenyl or 4-pyridyl). Compounds **76, 77, 78, 79** and **80** were found to cause significant increase in revertant colonies when compared with the solvent control whereas compound **79** showed most potent mutagenic activity against

Salmonella typhimurium TA98 and less mutagenic activity for TA100 (Tuylu et al., 2007). Kamal et al., in 2012 reported the G_2/M arrest and apoptotic inducing activities of chalcone-amidobenzothiazole conjugates. Compounds **(81, 82)** exhibited potent activity with an IC_{50} values in the range of 0.85-3.3 mM against different cancer cell lines. The compounds **(81)** (IC_{50} ¼ 3.5 mM) and **(82)** (IC_{50} ¼ 5.2 mM) inhibited microtubule assembly at both the molecular and cellular levels in A549 cells in addition to the tubulin polymerization (Kamal et al., 2012).

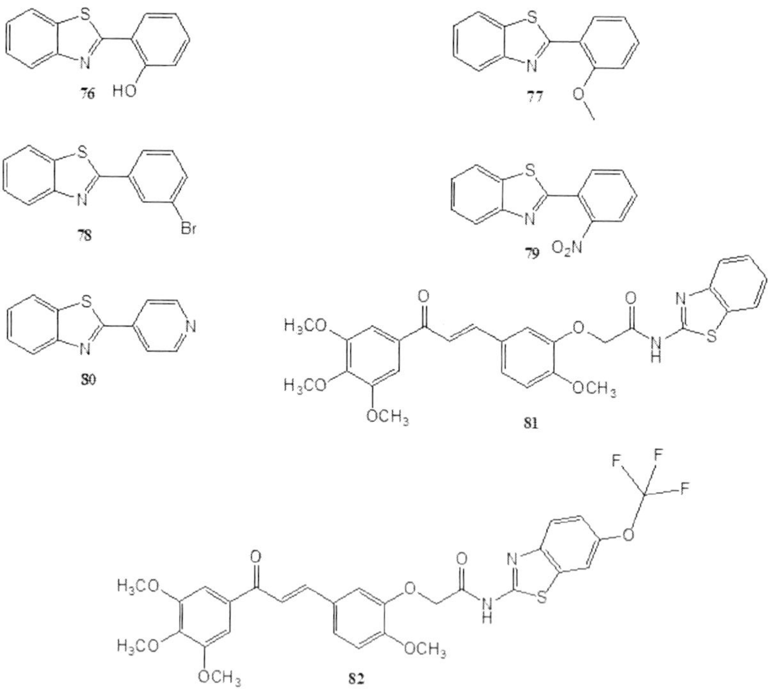

Figure 14a. Chemical structure of benzothiazole derivatives as replication and mitosis inhibitors.

2-Phenylimidazo (*2,1-b*) benzothiazole derivatives **(83-86)** have been reported to exhibit good antiproliferative activity, with GI_{50} values in the range 0.19-83.1 mM. Compound **(86)** showed potent anticancer efficacy against 60 human cancer cell lines, with a mean GI_{50} value of 0.88 mM. Also same compound induced cell-cycle arrest in the G_2/M phase and inhibited tubulin polymerization followed by activation of caspase-3 and apoptosis. (Kamal et al., 2012). Ashraf et al., synthesized a series of colchicine site binding tubulin inhibitors. They structurally modified the combretastatin A-4 (CA4) pharmacophore using benzothiazole moiety and evaluated their antiproliferative activity against selected cancer cell lines. The most potent compounds **87a** and **87b** demonstrated an antiproliferative effect at par with CA4 (GI_{50} = 0.06 ± 0.001 µM and 0.04 ± 0.001 µM, respectively) against HeLa cells (human cervical cancer cell line). The molecular docking studies suggested the synthesized compounds to bind at the colchicine site of the tubulin. Presence of a methoxy group on C-6 position of benzothiazole moiety was found to be an essential structural feature for antiproliferative activity (Ashraf et al., 2016). Hegde and his group investigated the ability of a benzothiazole derivative (**88**) to induce DNA damage and to inhibit cell proliferation in different cancer models. Compound **88** was found to cause significant G_2/M arrest along with deregulation of many cell cycle associated proteins such as CDK1, BCL2 and their phosphorylated form, CyclinB1, CDC25c etc. It also decreased levels of mitochondrial membrane potential and activation of apoptosis and inhibited tumor growth in mice without significant side effects (Hegde et al., 2017).

Shaik et al., in 2017 prepared and evaluated the potential of 2-anilinopyridinyl-benzothiazole Schiff bases to inhibit tubulin polymerization apart from inducing apoptosis. The compounds in the series were rationally designed based on the results of molecular modeling studies of the substituted compounds obtained by clubbing 2-trimethoxy-anilinopyridine and benzothiazole moieties via an imine linkage and by docking into the colchicine binding site of tubulin. The Schiff bases were obtained in good yields using convenient synthetic route. Though, most of the synthesized conjugates showed appreciable cytotoxicity against the various cancer cell lines but the Schiff base with trimethoxy group on benzothiazole moiety, **89** exhibited good antiproliferative activity with a GI_{50} value of 3.8µM specifically against the cell line DU145. In agreement with the docking results, **89** exerted cytotoxicity by the disruption of the microtubule dynamics by inhibiting tubulin polymerization via effective binding into colchicine domain, comparable to E7010. Furthermore, **89** induced apoptosis as evidenced by biological studies like mitochondrial membrane potential, caspase-3, and Annexin V-FITC assays (Shaik et al., 2017).

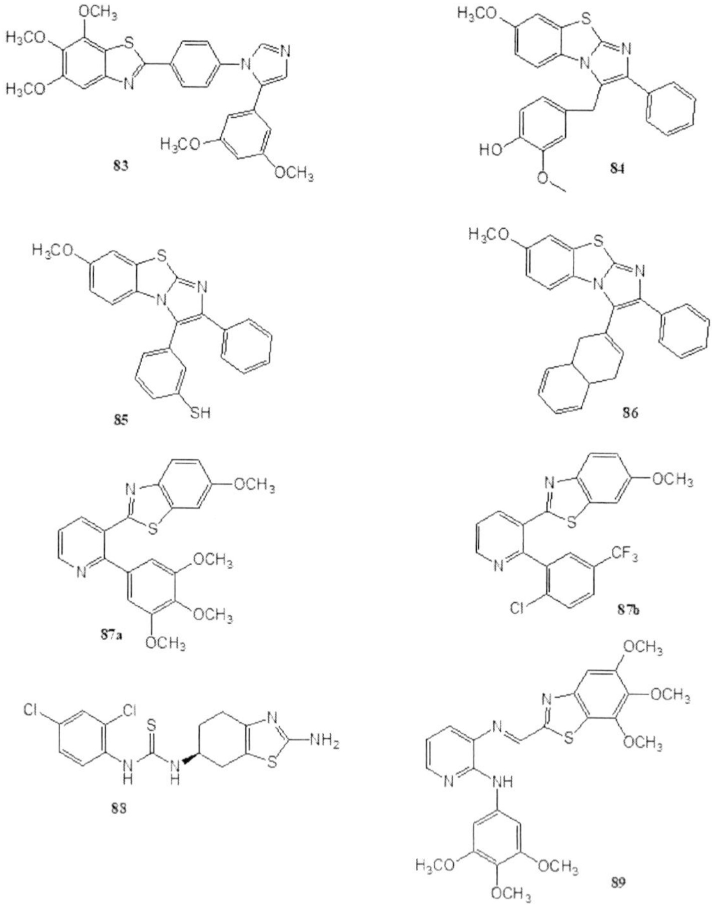

Figure 14b. Chemical structure of benzothiazole derivatives as replication and mitosis inhibitors.

2.3.6. Heat Shock Protein 90 (Hsp90) Inhibitors

As per the process of non oncogene addiction, cancer cells rely on the molecular chaperone, heat shock protein 70 (Hsp70), for survival and proliferation (Luo et al., 2009). Hsp70 is known to bind and stabilize many oncogenes and pro-survival "client" proteins, with the net effect of

Benzothiazole

suppressing multiple cell death pathways (Sherman & Gabai 2015; Calderwood & Gong, 2016; Srinivasan et al., 2017). Also, Hsp72 is commonly overexpressed in many cancers, where its levels correlate with poor patient prognosis, especially in cancers of the breast. Altogether, these observations indicate that Hsp70 family members may be good targets for the development of anticancer therapy (Murphy, 2015).

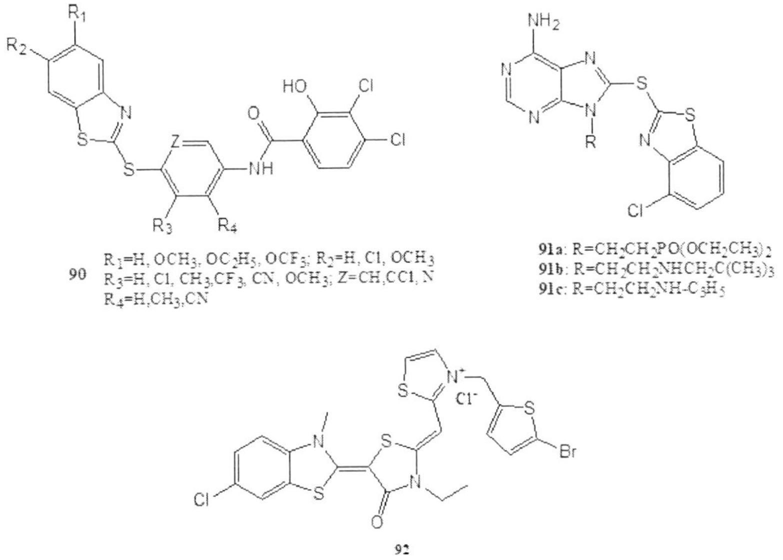

Figure 15. Chemical structure of benzothiazole derivatives as Hsp90 inhibitors.

Cathepsin D, a lysosomal aspartyl protease, has been implicated in the pathology of Alzheimer's breast and ovarian cancer in addition to AD. Taking a lead from this, benzothiazole analogues **90** were synthesized and screened as a cathepsin D inhibitor. It was observed that the heteroatom linker between the two rings can be either sulfur or oxygen, while substitution of the middle ring resulted in a slight increase in activity when a lipophilic substituent (chlorine, methyl, trifluoromethyl) is added ortho to the heteroatom linker. The overall potency of these analogues seems on track with the lipophilicity of the side-chain. (Dumas et al., 1999). Zhang et al., reported the synthesis of a series of benzo- and pyridinothiazolothiopurines as potent heat shock protein 90 inhibitors. Some of these compounds exhibited good aqueous solubility and oral bioavailability profiles in mice. Compounds **91a–c** were found to be the most potent inhibitors of tumour growth in an N87 human colon cancer xenograft model (Zhang et al., 2006).

Shao et al., in 2018 designed and developed Hsp70 inhibitors based on the JG-98 structure. Benzothiazole-rhodacyanines were shown to bind to the allosteric site on Hsp70, interrupting its binding to nucleotide-exchange factors (NEFs) to promote cell death in breast cancer cell lines. Compound **92** was found to reduce tumor burden in an MDA-MB-231 xenograft model (4 mg/kg; i.p.). It also activated caspase3/7 in MCF7 cells (Shao et al., 2018).

2.3.7. Human Germ Cell Alkaline Phosphatase (hGC-ALP) Inhibitors

The human germ cell alkaline phosphatase (hGC-ALP) is a known target for the development of prostate cancer chemotherapeutic agents. A series of some novel *N*-(benzo(*d*)thiazol-2-yl)-2-hydroxyquinoline- 4-carboxamides were synthesized following a four-step synthetic route using a range of substituted acetoacetanilides. The benzothiazole and quinoline fused bioactive compounds obtained in good yields using ecologically friendly catalytic systems were tested for antiprostate cancer efficacy through ALP inhibition potentials and using a range of cancer cell lines such as MCF-7 (Breast cancer), HCT-116 (Colon cancer), PC-3 & LNCaP (Prostate) and SK-HEP-1 (Liver cancer).

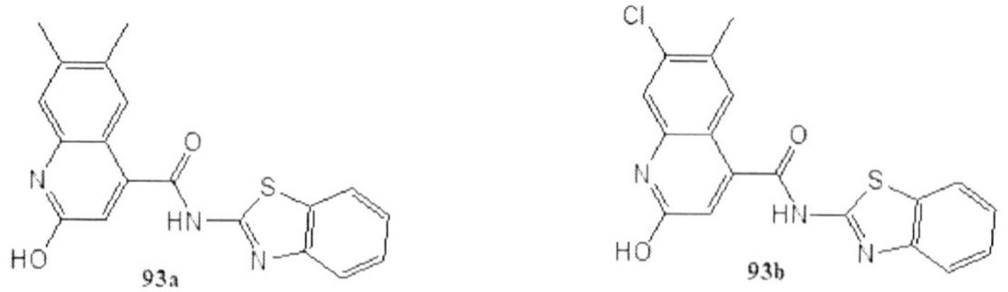

Figure 16. Chemical structure of benzothiazole derivatives as hGC-ALP inhibitors.

Among the synthesized compounds, **93a** and **93b** emerged as the selective antiproliferative agents against both the cancer cells (PC-3 and LNCaP). Based on the IC_{50} values, compounds **93a** (0.048 ± 0.002 μM) and **93b** (0.018 ± 0.002 μM) were observed to be more active than the standard bicalutamide (CDX) (1.3 ± 0.08 µM). Molecular docking studies of **93a** and **93b** revealed effective molecular interaction for ALP inhibition which were in agreement with the *in vitro* Human Germ Cell ALP assay (Bindu et al., 2019).

2.3.8. Cyclin-Dependent Kinase 2 (CDK2) Inhibitors

Cyclin-dependent kinases (CDKs) are well known for their vital roles in regulating cell divisions. They are considered as one of the most attractive targets for cancer therapy (Cheng et al., 2019). CDK2 is a member of the CDK family which is fundamental to the regulation of the cell cycle progression and is involved in the cell differentiation (Ying et al., 2018), and apoptosis (Faber & Chiles, 2007). Interestingly, CDK2 is frequently over-expressed in human tumors (Anscombe et al., 2015), while most normal tissues have low expression of CDK2. For these reasons, CDK2 has been emerged as one of the most promising therapeutic target for the discovery of highly efficient antitumor agents.

Diao and coworkers in 2019 designed and synthesized a series of novel pyrimidine-based benzothiazole hybrid derivatives as novel CDK2 Cyclin A2 inhibitors. Some of the synthesized hybrid molecules demonstrated quite remarkable antitumor activity against five cancer cell lines. One of the hybrids showed approximately potency with AZD5438 toward four cells including HeLa, HCT116, PC-3, and MDA-MB-231 with IC_{50} values of 0.45, 0.70, 0.92, 1.80 µM, respectively.

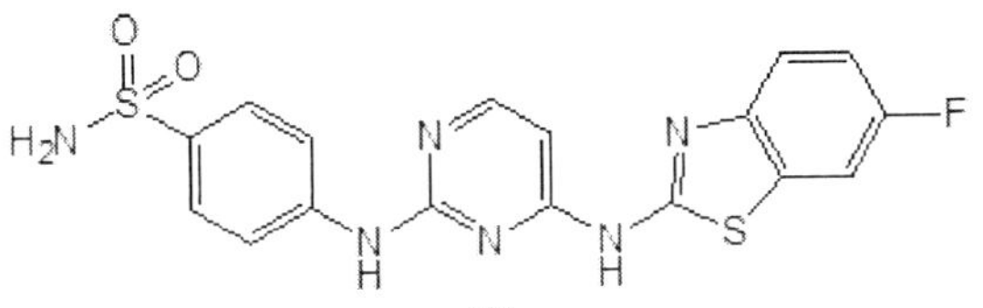

Figure 17. Chemical structure of a benzothiazole derivative as CDK2 inhibitors.

More interestingly, the most potent compound **94** also possessed promising CDK2/cyclin A2 inhibitory activities with IC_{50} values of 15.4 nM, which was almost 3-fold potent than positive control AZD5438.

The molecular docking studies indicated that the analogue bound efficiently with the CDK2 binding site. Further, the compound **94** was found to induce cell cycle arrest and apoptosis in a concentration-dependent manner. These observations suggest that pyrimidine-benzothiazole hybrids represent a new class of CDK2 inhibitors and well worth further investigation aiming to generate potential anticancer agents (Diao et al., 2019).

2.3.9 Inhibitors of Thioredoxin Signaling System

Thioredoxin/thioredoxin reductase system has the potential to act as target for anticancer drug and a small molecule inhibitor of this system is in clinical development. Lion et al., evaluated *in vitro* antitumour properties of a new series of fluorinated benzothiazole substituted- 4-hydroxycyclohexa-2,5-dienones (compounds **95a–c**). The new compounds were found to be of comparable activity as compared with the nonfluorinated precursors, in terms of antiproliferative activity in sensitive human cancer cell lines and inhibitory activity against the thioredoxin signaling system. Most potent antiproliferative activity was shown by the 5-fluoro analogue (**95a**) (Lion et al., 2006).

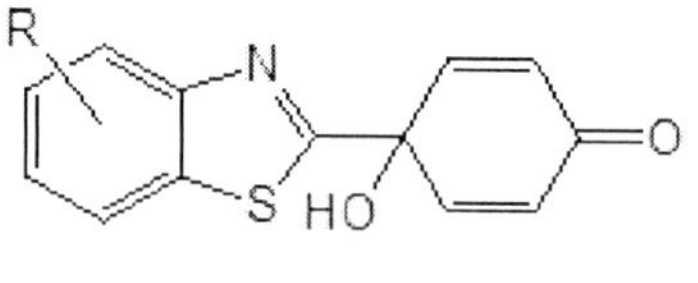

95a: R=5-F
95b: R=6-F
95c: R=4-F

Figure 18. Chemical structure of benzothiazole derivatives as thioredoxin signaling inhibitors.

2.3.10. Other Anticancer Benzothiazoles

Europium(III)- and terbium(III)-2-thioacetate benzothiazole complexes were synthesized by Hussein et al., in 2010 These complexes (**96a** and **96b**) showed strong binding affinity to calf thymus DNA using fluorometric and electronic absorption spectroscopy. Their antitumor effect was evident by the down-regulation of VEGF receptor type-2 (Flk-1). Moreover, the synthesized complexes exhibited significant cytotoxic activity against HepG2 and MCF7 cell lines and exhibited significant anticancer activity, when compared to the standard drug, cisplatin (Hussein et al., 2012).

Novel derivatives of *N*-alkylbromo-benzothiazoles were synthesized and evaluated for their anticancer potency by Gill et al., in 2013. Most of the compounds showed significant cytotoxic activity. However, compound **97**, (3-bromo-propyl)-(6-methoxy-benzothiazol-2-yl)- amine was found to be the most promising anticancer agent against the PC-3 (IC_{50} = 0.6 µM), THP-1 (IC_{50} = 3 µM), and Caco-2 cell lines (IC_{50} = 9.9 µM), respectively (Gill et al., 2013). Seenaiah and coworkers synthesized pyrimidinyl benzothiazoles linked by thio, methylthio, and amino moieties and studied their cytotoxic activities. The compounds were subjected to MTT assay on A549 lung adenocarcinoma cells to determine growth inhibitory/cytotoxic capability. The amino-linked pyrimidinyl bis-benzothiazole **98** exhibited cytotoxic activity on A549 cells with IC_{50} value of 10.5 µM (Seenaiah et al., 2014).

N-(6-Substitutedbenzothiazol-2-yl)-2-((5-((3-methoxyphenoxy)-methyl)-1,3,4-oxadiazol-2-yl) thio) acetamide derivatives were synthesized by Kaya et al.,. The synthesized compounds were screened for their anti-proliferative activity against two selected human tumor cell lines, A549 lung, MCF7 breast cancer cell line and mouse embryo fibroblast cell line, NIH/3T3 as healthy cell line. Among the compounds evaluated, compound **99** bearing 1,3,4-oxadiazole ring and 6-methoxy benzothiazole moiety exhibited the highest inhibitory activity against A549 and MCF-7 tumor cell lines in contrast to NIH/3T3 cell line (Kaya et al., 2017).

Novel berberine–benzothiazole conjugates were synthesized by Mistry et al. The synthesized compounds were screened for antioxidant potency using the DPPH and ABTS bioassays and for their *in vitro* anticancer activities against HeLa, CaSki (cervical cancer), and SK-OV-3 (ovarian cancer) cell lines using the SRB bioassay. The final compounds demonstrated significant antioxidant potency with IC_{50} levels of 13.03–24.50 µg/mL and 4.958–7.570 µg/mL in the DPPH and ABTS radical scavenging bioassays, respectively. Compound **100a**, with a methoxy functional group and compound **100b**, with a cyano functional group displayed maximum DPPH and ABTS radical scavenging activities.

Moreover, compound **100b** displayed the highest potency against all cancer cell lines, with IC_{50} value of 5.474, 5.311, and 32.61 μg/mL against the HeLa, CaSki, and SK-OV-3 cell lines, respectively (Mistry et al., 2017).

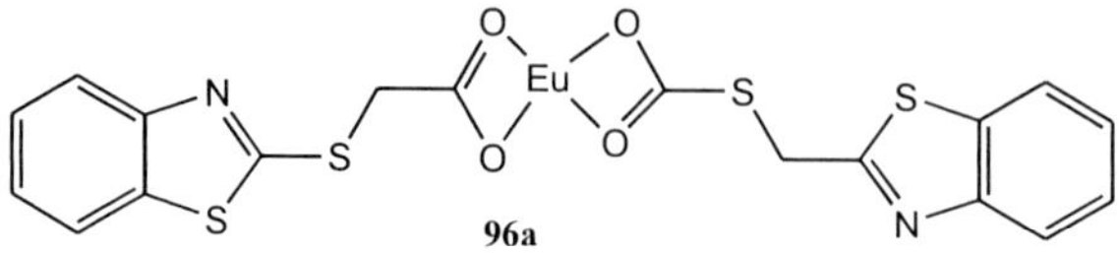

96a

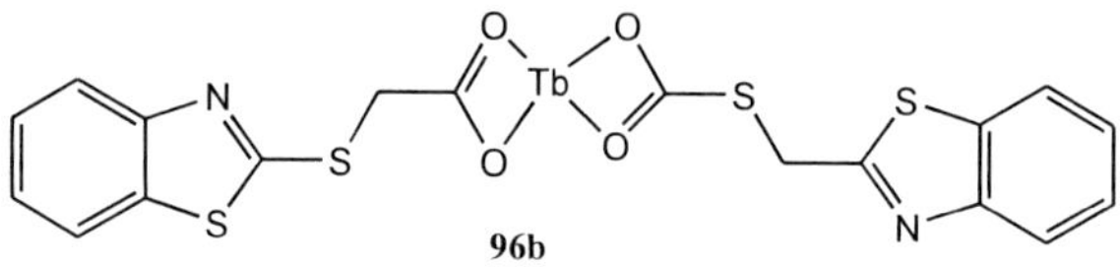

96b

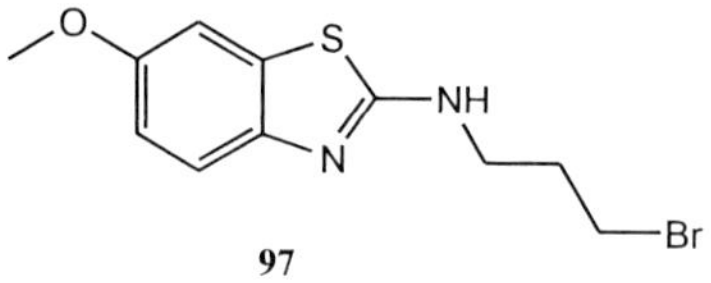

97

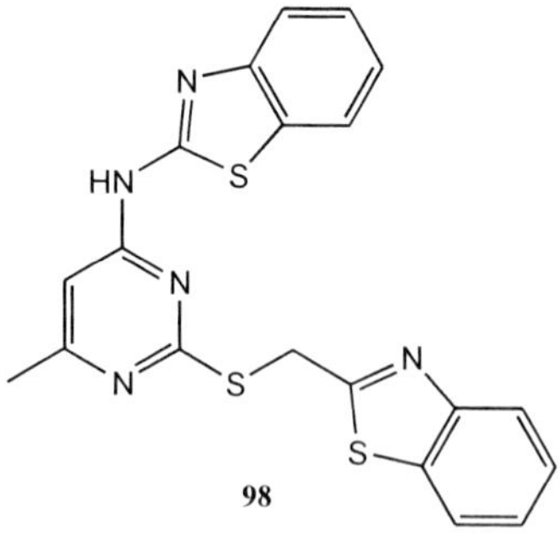

98

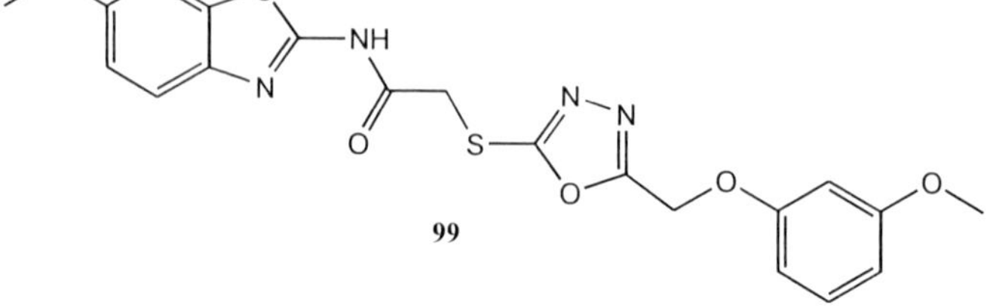

99

Figure 19. Chemical structure of benzothiazole derivatives as anticancer agents with miscellaneous mechanism of action

Cindric et al., designed and synthesized novel 2-imidazolinyl substituted benzo(*b*)thieno-2-carboxamides bearing benzothiazole subunit as potential antiproliferative agents. Their antiproliferative activities were evaluated on human cancer cell lines as well as on normal fibroblasts *in vitro*. Results revealed that most of the tested compounds showed moderate antiproliferative activities while cytotoxicity on normal fibroblasts was lower in comparison to the standard, 5-fluorouracil. Compound **101** emerged as the most potent against HeLa cells with a corresponding IC_{50} = 1.16 µM (Cindric et al., 2017).

A series of novel 4- and 5-substituted methylsulfonyl benzothiazole derivatives were synthesized by Lad et al., The synthesized compounds were screened for anticancer activity in cervical cancer (HeLa cell lines). Among the tested compounds, **102a** and **102b** showed significantly reduced cell growth with GI_{50} values ≤0.1 µM (Lad et al., 2017). Gong et al., investigated a new class of arylbenzothiazoles as potent, isoform-specific inhibitors of lysophosphatidic acid acyl transferase-β (LPAAT-β). Compound 103 inhibited human LPAAT-β with an IC_{50} value of 0.006 µM. Further these LPAAT-β inhibitors proved to be weakly antiproliferative against MCF-7 and DU145 cancer cell lines. Compound **103** showed moderate cytotoxicity with an IC_{50} value of 20 µM against DU145 cell line (Gong et al., 2004).

Modi et al., in 2019 selected *N*-(4-(benzo(*d*)thiazol-2-yl) phenyl)-5-chloro-2-methoxybenzamide **(104)** among the series of ten benzothiazole bearing amide derivatives and investigated its mechanism of the apoptotic pathway on cervical cancer cell lines. Compound **103** was demonstrated to generate ROS and DNA damage in SiHa and C-33A cells. The induction of apoptosis in SiHa cells is associated with increased nuclear expression of the tumor suppressor protein, TP53. The shift in BAX/BCL-2 ratio, increased expression of caspase-3 and cleaved Poly (ADP-ribose) polymerase-1 favour apoptotic signal in SiHa. In silico studies indicated that **104** inhibits the formation of E6/E6AP/P53 complex and thereby prevent degradation of TP53 and p53 mediated intrinsic apoptosis in SiHA. Thus, authors suggested that treatment of **104** leads to p53 and caspase dependent apoptosis in HPV 16 infected SiHa cells (Modi et al., 2019).

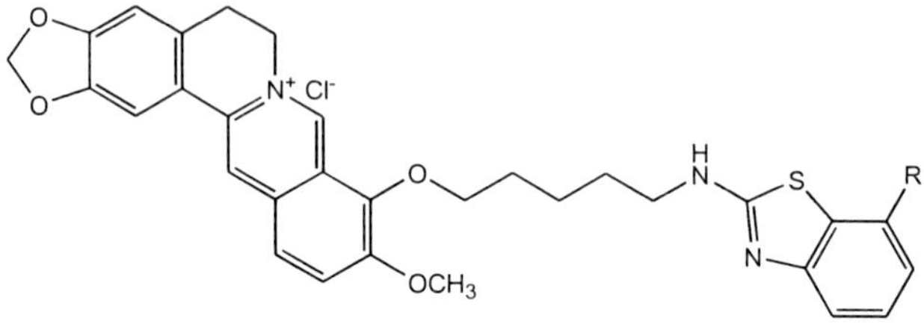

100a: R= $-OCH_3$
100b:R=-CN

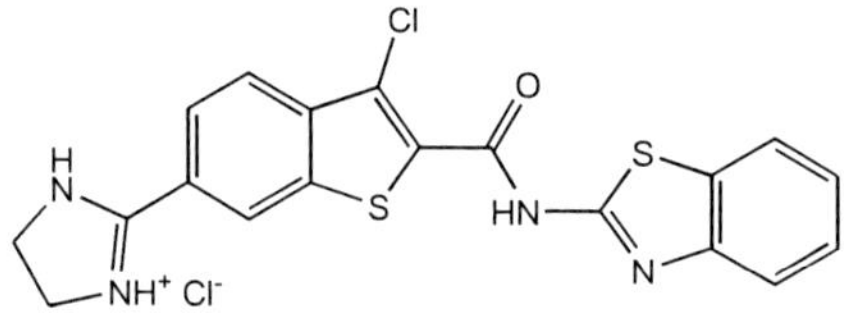

101

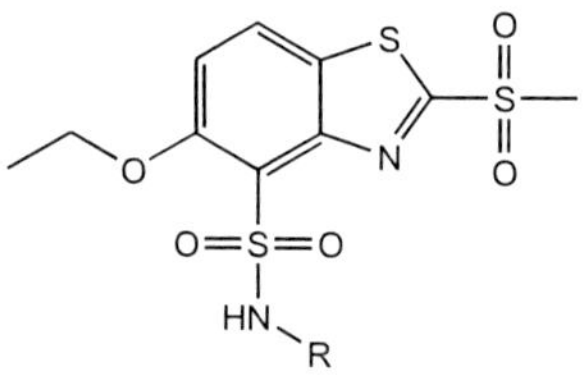

102a:R=p-C_6H_4-NO_2
102b:R=-$C(CH_3)_3$

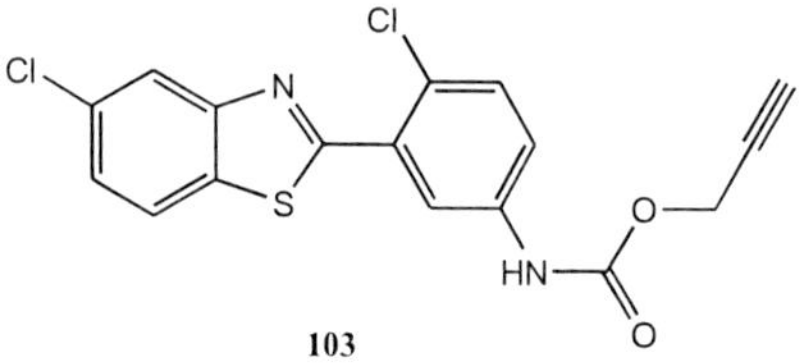

103

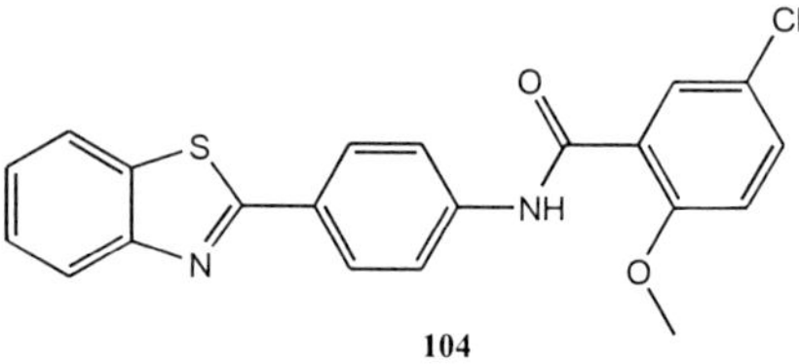

104

Figure 20. Chemical structure of benzothiazole derivatives as anticancer agents with miscellaneous mechanism of action

Peroxisome Proliferator-Activated Receptors (PPARs) constitute promising therapeutic targets for the development of medicinal compounds. Interestingly, the reduced activation of PPARs seems to have a positive impact on the growth and viability of cancer cells in multiple preclinical tumor models. Ammazzalorso and coworkers in 2019 reported the synthesis of some novel benzothiazole based amides obtained by structural modification in *N*-acylsulfonamides and evaluated their antiproliferative and antiPPARs activity in paraganglioma (PTJ64i and PTJ86i), pancreatic (AsPC-1 and Capan-2) and colorectal cancer cell lines (HT-29 and SW480). The synthesized compounds showed a moderate PPARα antagonistic activity (22-57%) in transactivation assay. However, in cellular assays they exhibited cytotoxicity in pancreatic, colorectal and paraganglioma cancer cells overexpressing PPARα. The most potent compound **105** showed the most remarkable inhibition of viability (>90%) in two paraganglioma cell lines, with IC$_{50}$ values in the low micromolar range. In addition, **105** markedly impaired colony formation capacity in the same cells. The results of this study showcase a relevant anti-proliferative potential of compound **105**, which appears particularly effective in paraganglioma, a rare tumor poorly responsive to chemotherapy (Ammazzalorso et al., 2019).

Liu et al., used a molecular hybridization strategy to design and synthesize some new pyrazole with benzo(*d*)thiazoles containing aminoguanidine structural units as anticancer agents. The 2-((1-(6-alkoxybenzo(*d*) thiazol-2-yl)-3-phenyl-1*H*-pyrazol-4-yl) methylene) hydrazinecarboximidamide derivatives were evaluated for their cytotoxixity and apoptosis promoting effects. The benzothiazole ring was substituted with varied functionalities to study their effect on antiproliferative activity. Amongst the synthesized compounds, (*E*)-2-((1-(6-((4-fluorobenzyl)oxy) benzo(*d*)thiazol-2-yl)-3-phenyl-1H-pyrazol-4-yl) methylene) hydrazine carboximidamide (**106**) emeged as the most potent compound, with IC$_{50}$ values of 2.41 µM, 2.23 µM, 3.75 µM and 2.31 µM *in vitro* anti-proliferative activity testing against triple-negative breast cancer cell line MDA-MB-231, non-triple-negative breast cancer MCF-7 cells, and human hepatocarcinoma HepG2 cells, and SMMC-7721 cells, respectively. Its activity particularly against MDA-MB-231 breast cancer cell line was found to be at par with reference anticancer drug doxorubicin. The results of flow cytometry mechanistic studies revealed the compound **106** to induce apoptosis of MDA-MB-231 cancer cells by downregulating Bcl-2 and upregulating Bax protein levels in a concentration-dependent manner. The structure activity relationship (SAR) studies showed that presence of fluoro substitution on the phenoxy ring attached at the 6th position of benzothiazole ring system is associated with better antitumor activity than chloro substitution. Further, the position of F-substituion on the phenoxy ring was found to affect the activity in the following order *p*-F>*m*-F>*o*-F (Liu et al., 2019).

A click synthesis approach in the presence and absence of the Cu(I) catalyst was used by Aouad et al., to design and synthesize a library of novel regioselective 1,4-di and 1,4,5-trisubstituted-1,2,3-triazole based benzothiazole-piperazine conjugates. Few of the synthesized 1,2,3-triazole hybrids were endowed with different heterocyclic moiety such as 1,2,4-triazole, benzothiazole, isatin and/or benzimidazole etc. 1,2,3-triazole hybrids showed moderate to potent antiproliferative activity against four selected human breast (MCF7, T47D) and human colon cancer cell lines (HCT116 and Caco2). The most potent compound **107**; 1-(4-(Benzothiazol-2-yl)piperazin-1-yl)-2-(4-(3-hydroxypropyl)-1H-1,2,3-triazol-1-yl)ethanone showed IC_{50} of 38 µM and 33 µM against the T47D and MCF7 cell lines, and IC_{50} of 48 µM and 42 µM against the HCT116 and Caco2. Based on the cell proliferation assay results, it can be inferred that hybrid molecules tethering ethyl ester substitution on the triazole moiety lack potent anticancer activity while aryl and alkyl substituted triazole is favored for imparting good anticancer activity (Aouad et al., 2018). The anticancer effects of 2-((1S,2S)-2-((E)-4-nitrostyryl)cyclopent-3-en-1-yl)benzo(d)thiazole and 2-((1S,2S)-2-((E)-4-fluorostyryl) cyclopent-3-en-1-yl)benzo(d)thiazole containing 2-substituted benzothiazoles were investigated against adenocarcinoma cancer cells (PANC-1) at various concentrations (5, 25, 50. 75 and 100 µM) by Uremis et al., They also determined the enzymatic antioxidant activity (superoxide dismutase (SOD), glutathione peroxidase (GPx)) and total antioxidant capacity (TAC) of the synthesized compounds. Authors showed the synthesized benzothiazole derivartives to exert antiproliferative effects against PANC-1 cells and to reduce cell viability by (i) inducing apoptosis of pancreatic cancer cells and (ii) inhibition of of SOD, GPx activity and by reducing TAC. The most promising compounds 2-((1S,2S)-2-((E)-4- nitrostyryl) cyclopent-3-en-1-yl)benzo(d)thiazole (**108a**) and 2-((1S,2S)-2-((E)-4-florostyryl)cyclopent-3-en-1-yl)benzo(d)thiazole (**108b**) displayed dose-dependent anti proliferative action on pancreatic cancer cell lines (IC_{50} = 27±0.24 µM and 35±0.51 µM, respectively). Infact, both the compounds were better than the positive control gemcitabine (IC_{50} of 52±0.72 µM). These findings suggest that the synthesized benzothiazole compounds may be considered as a potential therapeutic drug against human PANC-1 cancer cells (Uremis et al., 2017).

Mohammed and his group prepared a new series of cytotoxic benzothiazoles and tested their activity against human breast cancer MCF-7 cell line in comparison to cisplatin. Some of the synthesized compounds were found to possess moderate to good cytotoxicity but compound **109** (4-(2-(1-phenyl-*1H*-pyrazol-5*(4H)*-one-3-yl)diazen-1-yl)-5-oxo-3-phenyl-thiazo-lidin-2-ylidene-2-(1,3)benzothiazol-2-yl-acetonitrile), was noted to be more potent than cisplatin, with IC_{50} values of 5.15 μm with respect to 13.33 μm of cisplatin. In general, compounds obtained by coupling with aldehydes at C-4 of the thiazolidine ring and bearing hydroxyl group exhibited better cytotoxic activity but the compound **109** prepared by coupling a diazotized amine showed the best activity and emerged as the lead molecule. It was shown to regulate free radicals production, by increasing the activity of SOD and depletion of intracellular GSH, catalase, and GPx activities, accordingly, the high production of H_2O_2, nitric oxide, and other free radicals that caused tumor cell death as monitored by reduction in the synthesis of protein and nucleic acids (Mohamed et al., 2017).

The *N*-Mannich basees of berberine (an isoquinoline alkaloid) bearing substituted 2-aminobenzothiazole moieties were efficiently synthesized by Mistry et al., with an aim to improve the antioxidant and cytotoxic activities of pure berberine. The structurally modified analogues of berberin bearing benzothiazole ring showed significantly increased *in vitro* antioxidant activity and good growth inhibition of cervical cancer cell lines (HeLa and CaSki), an ovarian cancer cell line (SK-OV-3) and human renal cancer cell line (Caki-2). Compounds bearing a –OCH₃ substituent (**110a**; 12-(1-(6-Methoxy-2-aminobenzothiazole-1-ylmethyl)-berberrubine), -COOH group (**110b**; 12-(1-(2-Aminobenzothiazole-6-carboxylic acid-1-ylmethyl)-berberrubine), and a -CN substituent (**110c**; 12-(1-(6-Cyno-2-amino benzothiazole-1-ylmethyl)-berberrubine) displayed excellent radical scavenging activity in both DPPH and ABTS bioassays. However, the potent cytotoxicity of the modified analogues against the HeLa cell line was attributed to the presence of a 2-aminobenzothaizole moiety (**110d**; 12-(2-Aminobenzothiazole-1-yl methyl)-berberrubine) and its 6-chloro congener (**110e**; 12-(1-(6-Chloro-2-aminobenzothiazole-1-yl methyl)-berberrubine) on the berberine core, and the 6-cyano group (**110c**) on the benzothiazole ring revealed strong sensitivity for the CaSki cell line, though these compounds demonstrated diminished activity against the SK-OV-3 cell line. It was also observed that the compound with a 2-aminobenzo thaizole moiety (**110d**), compound with methoxy functional group (**110a**) and compound with cyano group exhibited the most significant cytotoxicity effect in Caki-2 cell line (Mistry et al., 2017).

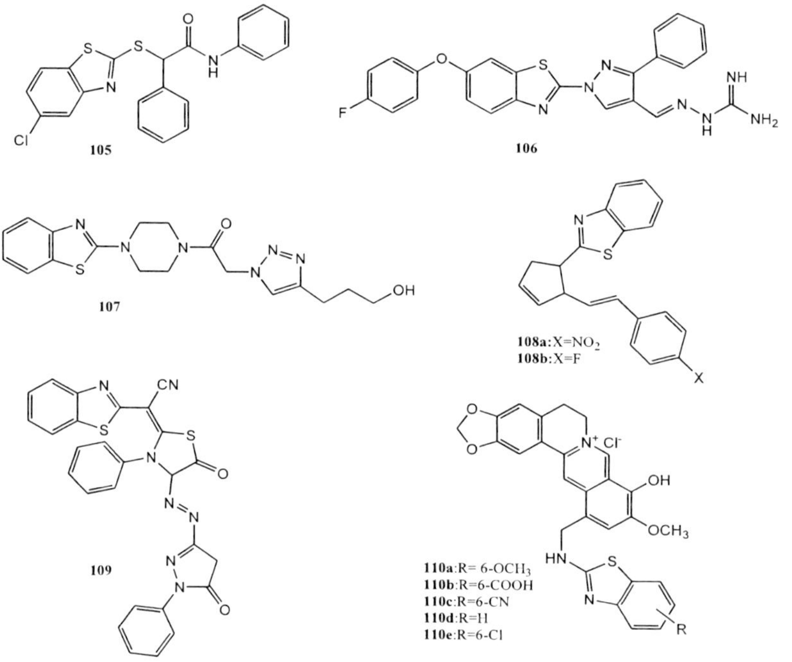

Figure 21. Chemical structure of benzothiazole derivatives as anticancer agents with miscellaneous mechanism of action

3. STRUCTURE-ACTIVITY RELATIONSHIP (SAR) OF BENZOTHIAZOLE SCAFFOLD WITH RESPECT TO ANTICANCER ACTIVITY

Benzothiazole derivatives exhibit varied spectrum of anticancer activity because of the benzothiazole moiety. This core is essential for exhibiting potent anticancer activity. SAR studies indicate that most of the benzothiazoles are substituted at 2nd position, 6th position, and 2nd and 6th

positions. Nature of these substituents or chemical functionality especially at 2nd position of benzothiazole ring can result in increase or decrease in activity. Based on the detailed analysis of identified potent anticancer benzothiazoles, some structural features were identified that could potentiate the anticancer activity of benzothiazole core nucleus by improving their physicochemical/ pharmacokinetc properties or selectivity for the cancerous cells. The important structural features along with their positions on benzothiazole ring are shown in Figure 22.

Following SAR is proposed for the benzothiazole derivatives as anticancer agents;

1. Presence of carboxamido/ureido connected to halogen substituted aromatic/cycloalkyl ring at 2nd position of benzothiazole ring imparts anti-Raf activity.
2. Presence of pyridylamide linked through ether linkage at 6th position of benzothiazole ring enhances Raf inhibitory activity.
3. Presence of halogens at 4,5,6 or 7th positions of benzothiazole and a *p*-substituted phenyl ring at 2nd position is associated with tyrosine kinase inhibitory activity.
4. A hydrazine moiety at 2nd position of benzothiazole is responsible for activity against cervical cancer and kidney fibroblast.
5. Incorporation of substituted benzyloxy phenyl ring at 2nd position of benzothiazole ring exhibits VEGFR inhibitory activity.
6. Presence of amino, hydroxyl and chloro groups in molecule is crucial for cytotoxicity.
7. Presence of a halogen in phenyl ring of benzothiazole enhances the anticancer spectrum.
8. Coupling of heterocyclic rings such as pyridine, thiazole, imidazole etc preferably at 2nd position were found to have significant effect on anticancer activity.
9. Coupling of bicyclic nitrogenous compounds through sulfur atom at 6th position increases their ability to inhibit c-Met.
10. Substitution with sulfonamidoaryl group at 6th position enhances PI3K inhibitory activity.

11. Presence of halogen at 5th position with substituted aniline group or hydroxyl group at 6th position along with substituted benzoyl group increases CYP450 inhibition.
12. Anticancer activity by inhibition of topoisomerase can be enhanced by increasing the number of hydroxyl groups on phenyl carboxamide attached to the 2nd position of benzothiazole.
13. Substitution at 3rd position of 2- phenyl ring with alkyl, amino or halogen atom leads to increase in the cytotoxicity.
14. Bulky substitution at C-4 of 2-phenyl ring potentiates the anticancer activity.

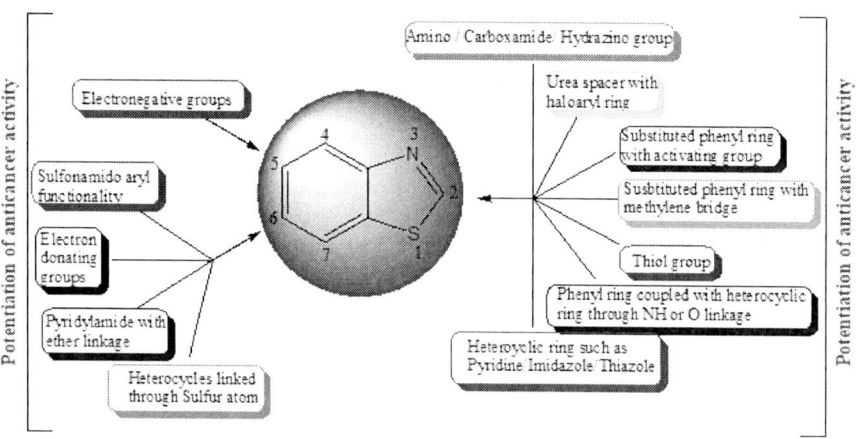

Figure 22. Structural features which enhances the anticancer activity of benzothiazole pharmacophore.

CONCLUSION

Benzothiazole is a versatile pharmacophore possessing an array of biological activities. The information presented in this article highlights the importance of naturally occurring benzothiazoles, strategies to synthesise benzothiazoles in good yield, patents granted to benzothiazole exhibiting anticancer activity, molecular targets of benzothiazoles in cancer chemotherapy, chemical structures and potential of recently synthesized benzothiazoles along with the structure activity relationship (Figure 22) as a blue print for the further potentiation of their anticancer activity. In spite of the interesting and diverse biological properties of naturally occurring benzothiazoles, very little synthetic work has been carried out on these natural products to identify and develop pharmacologically useful therapeutic agents. There is a vast scope for the medicinal chemists to explore and to modify the natural products containing benzothiazole ring in order to improve their biological activity spectrum.

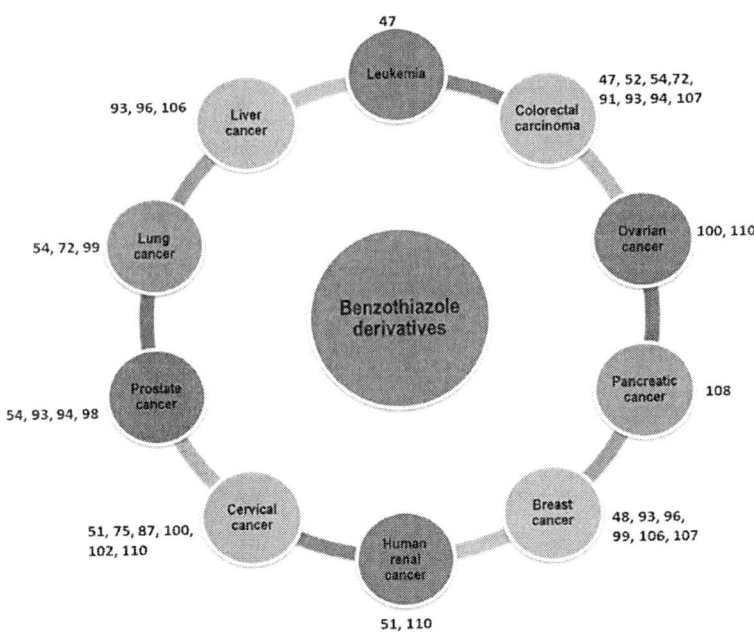

Figure 23. Usefulness of benzothiazole derivatives (with compound number) against various cancer cell lines.

The development of Phortress paved the way to focus on the development of similar benzothiazole analogues or conjugates. As a result, numerous synthetic benzothiazoles were synthesized in the past two decades and interestingly some of the benzothizoles derivatives owing to their mechanism of action at different molecular targets were reported to possess equivalent or better activity than the currently used chemotherapeutic agents. A number of benzothiazole derivatives reported in this article have shown to exhibit potent antiproliferative and cytotoxic activity against the most common cancers and is illustrated in Figure 23.

The superiority of these compounds in experimental studies over the clinically used anticancer agents makes them promising clinical drug candidates for the treatment of cancer.

However, there is a need to develop novel strategies for the synthesis of newer rationally designed benzothiazole hybrids, conjugates or prodrugs having good pharmacokinetic properties, safety, potency, specificity and selectivity towards the target to fight against the dreaded disease.

REFERENCES

Abdel-Aziz M, Matsuda K, Otsuka M, Uyeda M, Okawara T, Suzuki T. Inhibitory activities against topoisomerase I & II by polyhydroxybenzoyl amide derivatives and their structure-activity relationship. *Bioorg Med Chem Lett*. 2004;14:1669-1672.

Aiello S, Wells G, Stone EL, Kadri H, Bazzi R, Bell DR, Stevens MF, Matthews CS, Bradshaw TD, Westwell AD. Synthesis and biological properties of benzothiazole, benzoxazole, and chromen-4-one analogues of the potent antitumor agent 2-(3,4-dimethoxyphenyl)-5-fluoro-benzothiazole (PMX 610, NSC 721648). *J Med Chem*. 2008;51(16):5135-5139.

Alebrt E, Bacque E, Nemecek C, Ugolini A, Wentzler S. 6-*Triazolopyridazine sulfanyl benzothiazole derivatives as MET inhibitors*. US8546393 B2; 2013.

Alitalo K, Carmeliet P. Molecular mechanisms of lymphangiogenesis in health and disease. *Cancer Cell*. 2002;1:219-227.

Ammazzalorso A, De Lellis L, Florio R, Laghezza A, De Filippis B, Fantacuzzi M, Giampietro L, Maccallini C, Tortorella P, Veschi S, Loiodice F, Cama A, Amoroso R. Synthesis of novel benzothiazole amides: Evaluation of PPAR activity and anti-proliferative effects in paraganglioma, pancreatic and colorectal cancer cell lines. *Bioorg MedChem Lett.* 2019; 29(16):2302-2306.

Anjou K, Von Sydow E. The aroma of cranberries. II. *Vaccinium macrocarpon* Ait. *Acta Chem. Scand.* 1967; 21: 2076-2082.

Anscombe E, Meschini E, Moravidal R, et al., Identification and characterization of an irreversible inhibitor of CDK2. *Chem Biol.* 2015;22:1159-1164.

Aouad MR, Soliman MA, Alharbi MO, Bardaweel SK, Sahu PK, Ali AA, Messali M, Rezki N, Al-Soud YA. Design, Synthesis and anticancer screening of novel benzothiazole-piperazine-1,2,3-triazole hybrids. *Molecules.* 2018;23(11). pii: E2788. doi: 10.3390/molecules23112788.

Apuy JL, Inski DE, James JK. *Imidazobenzothiazoles as modulators of protein kinase.* WO 2010054058A1; 2010.

Arakawa R, Ito H, Takano A, Okumura M, Takahashi H, Takano H et al., Dopamine D2 receptor occupancy by perospirone: A positron emission tomography study in patients with schizophrenia and healthy subjects. *Psychopharmacol* (Berl). 2010;209:285–290.

Ashraf M, Shaik TB, Malik MS, Syed R, Mallipeddi PL, Vardhan MVPSV, Kamal A. Design and synthesis of cis-restricted benzimidazole and benzothiazole mimics of combretastatin A-4 as antimitotic agents with apoptosis inducing ability. *Bioorg Med ChemLett.* 2016; 26: 4527-4535.

Baba H, Yaoita Y, Kikuchi M. Sesquiterpenoids and lactone derivatives from *Ligularia dentate*. *Helv Chim Acta.* 2007; 90: 1028-1037.

Bacque E, Damour D, Nemececk C, et al., *Preparation of imidazo(1,2-a) pyrimidine derivatives as c-Met inhibitors and their pharmaceutical compositions.* WO007318; 2010.

Ben-Alloum A, Bakkasa S, Soufiaoui M. Nouvelle Voie de Synthese des 2-arylbenzothiazoles transfert d'electrons active par micro-ondes. *Tetrahedron Lett.* 1997;38(36): 6395-6396.

Bhagwat SS. *Preparation of imidazo (2,1-b)(1,3)benzothiazole derivatives for treatment of proliferative diseases.* WO056764; 2011.

Bhuva HA, Kini SG. Synthesis, anticancer activity and docking of some substituted benzothiazoles as tyrosine kinase inhibitors. *J Mol Graph Model.* 2010;29:32–37.

Bindu B, Vijayalakshmi S, Manikandan A. Discovery, synthesis and molecular substantiation of N-(benzo(*d*)thiazol-2-yl)-2-hydroxy-quinoline-4-carboxamides as anticancer agents. *Bioorg Chem*. 2019; 91:103171. doi: 10.1016/j.bioorg.2019.103171.

Boger DL. A convenient preparation of 2-substituted benzothiazoles. *J OrgChem*. 1978;43: 2296-2297.

Booker S, D'Angelo N, D'Amico DX et al., *Benzothiazole PI3 kinase modulators for cancer treatment*. US7928140 B2. 2011.

Bradshaw TD, Chua MS, Browne HL, Trapani V, Sausville EA Stevens MFG. In vitro evaluation of amino acid prodrugs of novel antitumour 2-(4-amino-3-methylphenyl) benzothiazoles. *British J Cancer*. 2002; 86:1348–1354.

Bradshaw TD, Stevens MF, Westwell AD. The discovery of the potent and selective antitumour agent 2-(4-amino-3-methylphenyl)benzothiazole (DF 203) and related compounds. *Curr Med Chem*. 2001;8:203-210.

Bradshaw TD, Westwell AD. The development of the antitumour benzothiazole prodrug, Phortress, as a clinical candidate. *Curr MedChem*. 2004;11(8):1009-1021.

Bruno RD, Njar VC. Targeting cytochrome p450 enzymes: a new approach in anticancer drug development. *Bioorg Med Chem*. 2007;15:5047-5060.

Calderwood SK, Gong J. Heat shock proteins promote cancer: It's a protection racket. *Trends Biochem Sci*. 2016; 41:311–323.

Carroll AR, Scheuer PJ. Kuanoniamines A, B, C, and D: pentacyclic alkaloids from a tunicate and its prosobranch mollusk predator *Chelynotus semperi*. *J Org Chem*. 1990; 55: 4426-4431.

Chen SW, Lee ZR, Li XY. Prediction of antifungal activity by support vector machine approach. *J Mol Struc Theochem*. 2005;731:73–81.

Chen Z, Su W, Shi X, Jiang L. *Method for preparation of 2-(1,3-disubstituted phenyl-4-pyrazolyl)benzothiazoles and application as antitumor agent*. CN101759694; 2010.

Cheng W, Yang Z, Wang S, Li Y, Wei H, Tian X, Kan Q, Recent development of CDK inhibitors: an overview of CDK/inhibitor co-crystal structures, *Eur J Med Chem*. 2019;164:615-639.

Chill L, Rudi A, Benayahu Y, Kashman Y. Violatinctamine, a new heterocyclic compound from the marine tunicate *Cystodytes cf. violatinctus*. *Tetrahedron Lett*. 2004; 45: 7925-7928.

Choi SJ, Park HJ, Lee SK, Kim SW, Han G, Choo HY. Solid phase combinatorial synthesis of benzothiazoles and evaluation of topoisomerase II inhibitory activity. *Bioorg Med Chem*. 2006;14:1229–1235.

Chua MS, Kashiyama E, Bradshaw TD, Stinson SF, Brantley E, Sausville EA et al., Role of CYP1A1 in modulation of antitumor properties of the novel agent 2-(4-amino-3- methylphenyl)benzothiazole (DF 203, NSC 674495) in human breast cancer cells. *Cancer Res*. 2000;60:5196–5203.

Cindrić M, Jambon S, Harej A, Depauw S, David-Cordonnier MH, Pavelik SK, Zamola GK, Hranjec M. Novel amidino substituted benzimidazole and benzothiazole benzo(*b*)thieno-2-carboxamides exert strong antiproliferative and DNA binding properties. *Eur J Med Chem*. 2017; 136: 468-479.

Damour D. *Preparation of 6-((6-alkyl or 6-cycloalkyl-triazolopyridazin-3-yl) sulfanyl)-1,3-benzothiazole derivatives as c-Met inhibitors and their pharmaceutical compositions*. WO121223; 2011.

Diao PC, Lin WY, Jian XE, Li YH, You WW, Zhao PL. Discovery of novel pyrimidine-based benzothiazole derivatives as potent cyclin-dependent kinase 2 inhibitors with anticancer activity. *Eur J MedChem*. 2019;179:196-207.

Ding LK, Jane Sutlive. The use of dithiazanine iodide in the treatment of multiple helminthiasis in Sarawak, Borneo. *Am J Trop Med Hyg*. 1960;9:503-505.

Ding Q, Huang XG, Wu J. Facile synthesis of benzothiazoles via cascade reactions of 2-iodoanilines, acid chlorides and Lawesson's reagent. *J Comb Chem*. 2009;11:1047–1049.

Dumas J, Brittelli D, Chen J, Dixon B, Hatoum-Mokdad H. Synthesis and structure, activity relationships of novel small molecule cathepsin D inhibitors. *Bioorg Med Chem Lett*. 1999; 9: 2531–2536.

El-Damasy AK, Cho NC, Nam G, Pae AN, Keum G. Discovery of a nanomolar multikinase inhibitor (KST016366): A new benzothiazole derivative with remarkable broad-spectrum antiproliferative activity. *Chem Med Chem.* 2016;11(15):1587-1595.

El-Damasy AK, Lee JH, Seo SH, Cho NC, Pae AN, Keum G. Design and synthesis of new potent anticancer benzothiazole amides and ureas featuring pyridylamide moiety and possessing dual B-Raf$^{(V600E)}$ and C-Raf kinase inhibitory activities. *Eur J Med Chem.* 2016;115:201-216.

Faber AC, Chiles TC. Inhibition of cyclin-dependent kinase-2 induces apoptosis in human diffuse large B-cell lymphomas. *Cell Cycle.* 2007;6:2982-2989.

Facchinetti V, Reis R, Gomes CRB, Vasconcelos TRA. Chemistry and biological activities of 1,3-benzothiazoles. *Mini Rev Org Chem.* 2012; 9: 44-53.

Feng Y, Lograsso P, Schroeter T, Yin Y. *Preparation of benzo(d)oxazoles and benzo(d)thiazoles as kinase inhibitor.* WO024903; 2010.

Ferrara N. VEGF as a therapeutic target in cancer. *Oncology.* 2005;69(Suppl 3):11-16.

Fichtner I, Monks A, Hose C, Stevens MF, Bradshaw TD. The experimental antitumor agents Phortress and doxorubicin are equiactive against human-derived breast carcinoma xenograft models. *Breast Cancer Res Treat.* 2004;87:97–107.

Fiehn O, Reemtsma T, Jekel M. Extraction and analysis of various benzothiazoles from industrial wastewater. *Anal Chim Acta.* 1994;295:297-305.

Furlan A, Colombo F, Kover A, Issaly N, Tintori C, Angeli L, Leroux V, Letard S, Amat M, Asses Y, Maigret B, Dubreuil P, Botta M, Dono R, Bosch J, Piccolo O, Passarella D, Maina F. Identification of new amino acid amides containing the imidazo(2,1-b)benzothiazol-2-ylphenyl moiety as inhibitors of tumorigenesis by oncogenic Met signaling. *Eur J Med Chem.* 2012;47:239-254.

Gabr MT, El-Gohary NS, El-Bendary ER, El-Kerdawy MM, Ni N. Synthesis, in vitro antitumor activity and molecular modeling studies of a new series of benzothiazole Schiff bases. *Chinese Chem Lett.* 2016;27:380–386.

Galmarini CM. Sagopilone, a microtubule stabilizer for the potential treatment of cancer. *Curr Opin Investig Drugs.* 2009;10(12):1359-1371.

Gan C, Zhou L, Zhao Z, Wang H. Benzothiazole Schiff-bases as potential imaging agents for β-amyloid plaques in Alzheimer's disease. *MedChem Res.* 2013;22:4069-4074.

Gentile A, Trusolino L, Comoglio PM. The Met tyrosine kinase receptor in development and cancer. *Cancer Metastasis Rev.* 2008;27:85-94.

Gibson P, Gill JH, Khan PA, Seargent PM, Martin, SW, Batman PA, Griffith J, Bradley C, Double JA, Bibby MC. Cytochrome p450 1B1 (CYP1B1) is overexpressed in human colon adenocarcinomas relative to normal colon: Implications for drug development. *Mol Cancer Ther.* 2003;2:527-534.

Gill RK, Singh G, Sharma A, Bedi PMS, Saxena AK. Synthesis, cytotoxic evaluation, and in silico studies of substituted N-alkylbromobenzothiazoles. *Med Chem Res.* 2013; 22: 4211-4222.

Gill RK, Rawal RK, Bariwal J. Recent advances in the chemistry and biology of Benzothiazoles. *Arch Pharm Chem Life Sci.* 2015; 348: 155–178.

Gong B, Hong F, Kohm C, et al., Synthesis and SAR of 2-arylbenzoxazoles, benzothiazoles and benzimidazoles as inhibitors of lysophosphatidic acid acyltransferase-β. *Bioorg Med Chem Lett.* 2004;14:1455-1459.

Gunawardana GP, Koehn FE, Lee AY, Clardy J, He HY, Faulkner DJ. Pyridoacridine alkaloids from deep-water marine sponges of the family Pachastrellidae: structure revision of dercitin and related compounds and correlation with the kuanoniamines. *J Org Chem.* 1992;57:1523-1526.

Gunawardana GP, Kohmoto S, Gunesakara SP, McConnel OJ, Koehn FE. Dercitine, a new biologically active acridine alkaloid from a deep water marine sponge, *Dercitus* sp. *J Am Chem Soc.* 1988;110:4856-4858.

Han SB, Nam NH, Hue VT, et al., *Preparation of hydroxamic acid compounds as histone deacetylase inhibitors and potent anticancer agents.* KR132657; 2012.

Hartley D, H. Kidd (eds), *The Agrochemical Handbook.* The Royal Society of Chemistry, Nottingham, England, 1987.

Hegde M, Vartak SV, Kavitha CV, Ananda H, Prasanna DS, Gopalakrishnan V, Choudhary B, Rangappa KS, Raghavan SC. A benzothiazole derivative (5g) induces DNA damage and potent G2/M arrest in cancer cells. *Sci Rep.* 2017;7(1):2533.

Henary M, Paranjpe, S, Owens, E. Synthesis and applications of benzothiazole containing cyanine dyes. *Heterocyclic Commun.* 2013;19(1): 1-11.

Hong SH, Hong IS, Han EJ, Lee GH. *Preparation of benzothiazole derivatives for the treatment of cancer.* KR128693;2013.

Hosokawa N, Naganawa H, Hamada M, Iinuma H, Takeuchi T, Tsuchiya KS, Hori M. New triene-ansamycins, thiazinotrienomycins F and G and a diene-ansamycin, benzoxazomycin. *J Antibiot* (Tokyo). 2000; 53: 886-894.

Hur Y, Lee HJ, Kim EK, et al., *Imide-containing benzothiazole derivatives and pharmaceutical composition comprising the same.* WO043002; 2013.

Hussein BH, Azab HA, el-Azab MF, el-Falouji AI. A novel anti-tumor agent, Ln(III) 2-thioacetate benzothiazole induces anti-angiogenic effect and cell death in cancer cell lines. *Eur J Med Chem.* 2012;51:99-109.

Hutchinson I, Bradshaw TD, Matthews CS, Stevens MF, Westwell AD. Antitumour benzothiazoles. Part 20: 3 ′ -cyano and 3 ′ -alkynylsubstituted 2-(4 ′ -aminophenyl)benzothiazoles as new potent and selective analogues. *Bioorg Med Chem Lett.* 2003;13:471–474.

Hutchinson I, Chua MS, Browne HL, Trapani V, Bradshaw TD, Westwell AD, Stevens MFG. Antitumor benzothiazoles. 14. Synthesis and in vitro biological properties of fluorinated 2-(4-aminophenyl) benzothiazoles. *J Med Chem.* 2001;44:1446–1455.

Hutchinson I, Jennings SA, Vishnuvajjala BR, Westwell AD, Stevens MF. Antitumor benzothiazoles. 16. Synthesis and pharmaceutical properties of antitumor 2-(4-aminophenyl) benzothiazole amino acid prodrugs. *J Med Chem.* 2002;45:744–747.

Ind PW, Laitinen L, Laursen L, Wenzel S, Wouters E, Deamer L, Nystrom P. Early clinical investigation of Viozan (sibenadet HCl), a novel D2 dopamine receptor, beta2-adrenoceptor agonist for the treatment of chronic obstructive pulmonary disease symptoms. *Respir Med.* 2003;97 Suppl A:S9-21.

Jackson PF, Maharoof USM, Leonard KA, et al., *Preparation of benzothiazolyl inhibitors of pro-matrix metalloproteinase activation.* WO162463; 2012.

Jain A, Sharma R, Gahalain N. Anxiety disorder: An overview. *Int J Pharmacy Life Sci.* 2010;1:396–409.

Kamal A, Kumar BA, Suresh P, et al., An efficient one-pot synthesis of benzothiazolo-4β-anilinopodophyllotoxin congeners: DNA topoisomerase-II inhibition and anticancer activity *Bioorg Med Chem Lett.* 2011;21:350-353.

Kamal A, Mallareddy A, Suresh P, et al., *Benzothiazole hybrids useful as anticancer agents and process for the preparation thereof.* INDE00270; 2011.

Kamal A, Mallareddy A, Suresh P, et al., *Benzothiazole hybrids useful as anticancer agents and process for the preparation thereof.* WO104875; 2012.

Kamal A, Rajesh VC, Reddy KS, et al., *Preparation of pyrrolo(2,1-c)(1,4) benzodiazepine-benzothiazole or -benzoxazole conjugates linked through piperazine moiety for use in treating cancer.* WO117882; 2011.

Kamal A, Reddy CR, Prabhakar S. *2-Phenyl benzothlazole linked imidazole compounds as potential anticancer agents.* INDE00392; 2014.

Kamal A, Reddy CR, Prabhakar S. *Imidazolylphenyl)benzothiazole derivatives as anticancer agents and process for the preparation thereof.* WO110959; 2012.

Kamal A, Srikanth YV, Naseer Ahmed Khan M, et al., *Benzothiazole derivatives as potential anticancer agents and apoptosis inducers and process for the preparation thereof.* WO111016; 2012.

Kamal A, Suresh MP, Shaik TB, Nayak VL, Kishor C, Shetti R, Rao NS, Tamboli JR, Ramakrishna S, Addlagatta A. Synthesis of chalconeamido benzothiazole conjugates as antimitotic and apoptotic inducing agents. *Bioorg Med Chem.* 2012;20:3480-3492.

Kamal A, Sultana F, Ramaiah MJ, Srikanth YV, Viswanath A, Kishor C, Sharma P, Pushpavalli SN, Addlagatta A, Pal-Bhadra M. 3- Substituted

2-phenylimidazo(2,1-b)benzothiazoles: synthesis, anticancer activity, and inhibition of tubulin polymerization. *Chem Med Chem*. 2012;7:292-300.

Kamal A, Faazil S, Ramaiah MJ, Ashraf M, Balakrishna M, Pushpavalli SN, Patel N, Pal-Bhadra M. Synthesis and study of benzothiazole conjugates in the control of cell proliferation by modulating Ras/MEK/ERK-dependent pathway in MCF-7 cells. *Bioorg Med Chem Lett*. 2013; 23:5733-5739.

Katritzky AR, Rees CW. *Comprehensive heterocyclic chemistry- Structure, reactions, synthesis and uses of heterocyclic compounds*. Pergamon Press Oxford, 1984; vol. 1, 143-152.

Kavallaris M. Microtubules and resistance to tubulin-binding agents. *Nat Rev Cancer*. 2010;10:194-204.

Kaya B, Hussin W, Yurttaş L, Turan-Zitouni G, Gençer HK, Baysal M, Karaduman AB, Kaplancıklı ZA. Design and synthesis of new 1,3,4-oxadiazole - benzothiazole and hydrazone derivatives as promising chemotherapeutic agents. *Drug Res* (Stuttg). 2017; 67:275-282.

Khokra SL, Arora K, Khan SA, Kaushik P, Saini R, Husain A. Synthesis, computational studies and anticonvulsant activity of novel benzothiazole coupled sulfonamide derivatives. *Iranian J Pharm Res*. 2019;18(1):1-15.

Klunk WE, Engler H, Nordberg A, Wang Y, Blomqvist G, Holt DP, Bergström M, Savitcheva I, Huang GF, Estrada S, Ausén B, Debnath ML, Barletta J, Price JC, Sandell J, Lopresti BJ, Wall A, Koivisto P, Antoni G, Mathis CA, Långström B. Imaging brain amyloid in Alzheimer's disease with Pittsburgh Compound-B. *Ann Neurol*. 2004;55(3):306-319.

Kodama T, Tanaka T, Kawamura T, et al., *Pr-set7 inhibitor and therapeutic and/or preventive agent for cancer or life style-related disease*. WO010715; 2011.

Kurtz JE, Ray-Coquard I. PI3 kinase inhibitors in the clinic: an update. *Anticancer Res*. 2012;32:2463-2470.

Lad NP, Manohar Y, Mascarenhas M, Pandit YB, Kulkarni MR, Sharma R, Salkar K, Suthar A, Pandit SS. Methylsulfonyl benzothiazoles (MSBT) derivatives: Search for new potential antimicrobial and anticancer agents. *Bio Med Chem Lett.* 2017; 27: 1319-1324.

Leong CO, Gaskell M, Martin EA, Heydon RT, Farmer PB, Bibby MC et al., Antitumour 2-(4-aminophenyl) benzothiazoles generate DNA adducts in sensitive tumour cells in vitro and in vivo. *Brit J Cancer.* 2003;88:470–477.

Leong CO, Suggitt M, Swaine DJ, Bibby MC, Stevens MFG, Bradshaw TD. In vitro, in vivo, and in silico analyses of the antitumor activity of 2-(4-amino-3-methylphenyl)-5- fluorobenzothiazoles. *Mol Cancer* Ther. 2004;3:1565–1575.

Li X, Liu X, Wang C, et al., *2-(4-Aminophenyl) benzothiazole-containing naphthalimide derivatives useful in treatment of cancers and their preparation.* CN103450176; 2013.

Li X, Ma L, Zhao J, et al., *Preparation of mercapto benzothiazole derivatives as antitumor agents.* CN103435574; 2013.

Li Y, Mei L, Li H, et al., *Predation of 2-methoxy-3-substituted-sulfonylamino-5-(2-acetamido-6-benzothiazole)-benzamide derivatives as antitumor agents.* CN103772317; 2014.

Lion CJ, Matthews CS, Wells G, Bradshaw TD, Stevens MF, Westwell AD. Antitumour properties of fluorinated benzothiazole substituted hydroxycyclohexa-2,5-dienones ('quinols'). *Bioorg Med Chem Lett.* 2006;16:5005–5008.

Liu C, Leftheris K, Lin J. *Benzothiazole and azabenzothiazole compounds useful as kinase inhibitors.* US20090118272 A1; 2009.

Liu DC, Gao MJ, Huo Q, Ma T, Wang Y, Wu CZ. Design, synthesis, and apoptosis-promoting effect evaluation of novel pyrazole with benzo(*d*)thiazole derivatives containing aminoguanidine units. *J Enzyme Inhib Med Chem.* 2019;34(1):829-837.

Luo J, Solimini NL, Elledge SJ. Principles of cancer therapy: oncogene and non-oncogene addiction. *Cell* 2009;136:823–837.

Mistry B, Patel RV, Keum YS, Kim DH. Evaluation of the biological potencies of newly synthesized berberine derivatives bearing benzothiazole moieties with substituted functionalities *J Saudi Chem Soc.* 2017; 21: 210-219.

Mistry BM, Shin HS, Keum YS, Pandurangan M, Kim DH, Moon SH, Kadam AA, Shinde SK, Modi A, Singh M, Gutti G, Shanker OR, Singh VK, Singh S, Singh SK, Pradhan S, Narayan G. Benzothiazole derivative bearing amide moiety induces p53-mediated apoptosis in HPV16 positive cervical cancer cells. *Invest New Drugs*. 2019. doi: 10.1007/s10637-019-00848-7.

Mohamed LW, Taher AT, Rady GS, Ali MM, Mahmoud AE. Synthesis and cytotoxic activity of certain benzothiazole derivatives against human MCF-7 cancer cell line. *Chem Biol Drug Des*. 2017;89(4):566-576.

Morton K. Convulsions following diamthazole (Asterol). *AMA Am J DisChild*. 1960;99(1):109.

Murphy ME. The HSP70 family and cancer. *Carcinogenesis*. 2013; 34:1181–1188.

Murray GI, Taylor MC, McFadyen MC, McKay JA, Greenlee WF, Burke MD, Melvin WT. Tumor-specific expression of cytochrome P450 CYP1B1. *Cancer Res*. 1997;57:3026-3031.

Murray GI. The role of cytochrome P450 in tumour development and progression and its potential in therapy. *J Pathol*. 2000;192:419-426.

Nagaraju B, Kovvuri J, Kumar CG, Routhu SR, Shareef MA, Kadagathur M, Adiyala PR, Alavala S, Nagesh N, Kamal A. Synthesis and biological evaluation of pyrazole linked benzothiazole-β-naphthol derivatives as topoisomerase I inhibitors with DNA binding ability. *Bioorg Med Chem*. 2019;27(5):708-720.

Nageswari A, Reddy PP. *Synthesis and spectral characterization of related compounds of riluzole an amyotrophic lateral sclerosis drug substance.* ARKIVOC. 2008;14:109–114.

Nelissen N, Van Laere K, Thurfjell L, Owenius R, Vandenbulcke M, Koole M, et al., Phase 1 study of the Pittsburgh compound B derivative 18Fflutemetamol in healthy volunteers and patients with probable Alzheimer disease. *J Nucl Med*. 2009;50:1251–1259.

Nemececk C, Ugolini A, Wentzler S. *Preparation of 6-(6-O-substitutedtriazolopyridazinesulfanyl)Benzothiazole and benzimidazole derivatives as c-Met inhibitors and their pharmaceutical compositions.* WO0089507; 2010.

Okaniwa M, Takagi T. *Preparation of benzothiazole derivatives as anticancer agents.* WO064722; 2010.

Okaniwa M, Takag T, Hirose M. *Benzothiazole derivatives as anticancer agents.* EP2358689 A1; 24 Aug 2011.

Okaniwa M, Takag T, Hirose M. *Benzothiazole compounds useful for Raf inhibition.* US8143258B2. 2009.

Ozen CK, Gulbas BT, Foto E, Yildiz I, Diril N, Aki E, Yalcin I. Benzothiazole derivatives as human DNA topoisomerase IIa Inhibitors. *Med Chem Res.* 22 (2013) 5798-5808.

Panicker B, Oehlen LJWM, Tarrant JG, et al., *Preparation of benzo(d)thiazole compounds as cytochrome P450 inhibitors and therapeutic uses thereof.* WO153192; 2011.

Patel RV. Synthesis and evaluation of antioxidant and cytotoxicity of the *N*-Mannich base of berberine bearing benzothiazole moieties. *Anticancer Agents Med Chem.* 2017;17(12):1652-1660.

Pinar A, Yurdakul P, Yildiz I, Temiz-Arpaci O, Acan NL, Aki-Sener E, Yalcin I. Some fused heterocyclic compounds as eukaryotic topoisomerase II inhibitors. *Biochem Biophys Res Commun.* 2004; 317:670-674.

Reddy VG, Reddy TS, Jadala C, Reddy MS, Sultana F, Akunuri R, Bhargava SK, Wlodkowic D, Srihari P, Kamal A. Pyrazolo-benzothiazole hybrids: Synthesis, anticancer properties and evaluation of antiangiogenic activity using in vitro VEGFR-2 kinase and *in vivo* transgenic zebrafish model. *Eur J Med Chem.* 2019;182:111609.

Reemtsma T, Fiehn O, Kalnowski G, Jekel M. Microbial transformations and biological effects of fungicide-derived benzothiazoles determined in industrial waste water. *Environ Sci Technol.* 1995;29:478-485.

Satyanarayana B, Saravanan M, Kumari KS, Lokamaheshwari DP, Sridhar Ch, Ravishankar R, Scheetz ME II, Carlson DG, Schinitsky MR. Frentizole, a novel immunosuppressive, and azathioprine: Their comparative effects on host resistance to *Pseudomonas aeruginosa, Candida albicans*, herpes simplex virus, and influenza (Ann Arbor) virus. *Infect Immun.* 1977;15:145–148.

Seenaiah D, Reddy PR, Reddy GM, Padmaja A, Padmavathi V, Krishna NS. Synthesis, antimicrobial and cytotoxic activities of pyrimidinyl benzoxazole, benzothiazole and benzimidazole. *Eur J Med Chem.* 2014;77:1-7.

Seifert RM, King Jr AD. Identification of some volatile constituents of *Aspergillus clavatus*. *J Agric Food Chem.* 1982;30:786-790.

Sensi P, Margalith P, Timbal MT. Rifomycin, a new antibiotic; preliminary report. *Farmaco Sci.* 1959;14:146-147.

Shahare HV, Talele GS. Designing of benzothiazole derivatives as promising EGFR tyrosine kinase inhibitors: a pharmacoinformatics study. *J Biomol Struct Dyn.* 2019:1-10. doi: 10.1080/07391102.2019.1604264.

Shaik TB, Hussaini SMA, Nayak VL, Sucharitha ML, Malik MS, Kamal A. Rational design and synthesis of 2-anilinopyridinyl-benzothiazole Schiff bases as antimitotic agents. *Bioorg Med Chem Lett.* 2017;27(11):2549-2558.

Shao H, Li X, Moses MA, Gilbert LA, Kalyanaraman C, Young ZT, Chernova M, Journey SN, Weissman JS, Hann B, Jacobson MP, Neckers L, Gestwicki JE. Exploration of benzothiazole-rhodacyanines as allosteric inhibitors of protein-protein interactions with heat shock protein 70 (Hsp70). *Med Chem.* 2018;61(14):6163–6177.

Sherman MY, Gabai VL. Hsp70 in cancer: back to the future. *Oncogene* 2015; 34:4153–4161.

Shi DF, Bradshaw TD, Wrigley S, et al., Antitumor benzothiazoles. 3. Synthesis of 2-(4-aminophenyl) benzothiazoles and evaluation of their activities against breast cancer cell lines *invitro and in vivo*. *J MedChem.* 1996;39:3375-3384.

Shin-ya K, Umeda Y, Chijiwa S, Furihata K, Furihata K, Hayakawa Y, Seto H. Mevashuntin, a novel metabolite produced by inhibition of the mevalonate pathway in *Streptomyces prunicolor*. *Tetrahedron Lett.* 2005;46:1273-1276.

Spadaro A, Frotscher M, Hartmann RW. Optimization of hydroxybenzothiazoles as novel potent and selective inhibitors of 17β-HSD1. *J Med Chem.* 2012;55:2469-2473.

Srinivasan SR, Cesa LC, Li X, Julien O, Zhuang M, Shao H, Chung J, Maillard I, Wells JA, Duckett CS, Gestwicki JE. Heat shock protein 70 (Hsp70) suppresses RIP1-dependent apoptotic and necroptotic cascades. *Mol Cancer Res.* 2017;16: 58–68.

Su TL, Chou TC, Lee TC. *Synthesis of 4H-benzo(d)pyrrolo(1,2-a)thiazoles and indolizino(6,7-b) indole derivatives as antitumor therapeutic agents.* US0178629; 2013.

Supuran CT. Carbonic anhydrases: Novel therapeutic applications for inhibitors and activators. *Nat Rev Drug Discov.* 2008;7:168–181.

Suzuki H, Shindo K, Ueno A, Miura T, Takei M, Sakakibara M, Fukamachi H, Tanaka J, Higa T. S1319: a novel beta2-andrenoceptor agonist from a marine sponge Dysidea sp. *Bioorg Med Chem Lett.* 1999; 9: 1361-1364.

Swinnen D, Jorand-Lebrun C, Grippi-Vallotton T, et al., *Fused bicyclic derivatives as PI3 kinase inhibitors and their preparation and use for the treatment of diseases.* WO100144; 2010.

Tanaka R, Kawamura T, Hirayama N. Structure of tiaramide. *Anal Sci.* 2007;23:105–106.

Tang JCO, Chan ASC, Lam KH, et al., *Preparation of methylcantharimide derivatives as antitumor agents.* US8344007; 2013.

Tasler S, Müller O, Wieber T, Herz T, Krauss R, Totzke F, Kubbutat MH, Schächtele C. *N*-substituted 2'-(aminoaryl) benzothiazoles as kinase inhibitors: hit identification and scaffold hopping. *Bioorg Med Chem Lett.* 2009;19:1349-1356.

Tasler S, Müller O, Wieber T, Herz T, Pegoraro S, Saeb W, Lang M, Krauss R, Totzke F, Zirrgiebel U, Ehlert JE, Kubbutat MH, Schächtele C. Substituted 2-arylbenzothiazoles as kinase inhibitors: hit-to-lead optimization. *Bioorg Med Chem.* 2009;17:6728-6737.

Tsai J, Lee JT, Wang W, Zhang J, Cho H, Mamo S, Bremer R, Gillette S, Kong J, Haass NK, Sproesser K, Li L, Smalley KS, Fong D, Zhu YL, Marimuthu A, Nguyen H, Lam B, Liu J, Cheung I, Rice J, Suzuki Y, Luu C, Settachatgul C, Shellooe R, Cantwell J, Kim SH, Schlessinger J, Zhang KY, West BL, Powell B, Habets G, Zhang C, Ibrahim PN, Hirth P, Artis DR, Herlyn M, Bollag G. Discovery of a selective inhibitor of oncogenic B-Raf kinase with potent antimelanoma activity. *Proc Natl Acad Sci. USA*. 2008;105(8):3041–3046.

Tuylu BA, Zeytinoglu HS, Isikdag I. Synthesis and mutagenicity of 2-aryl-substitute (o-hydroxy- m-bromo- o-methoxy- o-nitrophenyl or 4-pyridyl) benzothiazole derivatives on *Salmonella typhimurium* and human lymphocytes exposed *in vitro*. *Biologia*. 2007;62:626–632.

Uremis N, Uremis MM, Tolun FI, Ceylan M, Doganer A, Kurt AH. Synthesis of 2-substituted benzothiazole derivatives and their in vitro anticancer effects and antioxidant activities against pancreatic cancer cells. *Anticancer Res*. 2017;37(11):6381-6389.

Vinodhini C, Sravani MVND, Bhanuprakash M, Devi MS, Imran M, Osman MOA et al., Method development and validation of pramipexole dihydrochloride monohydrate in tablet dosage form by UV and visible spectrophotometric methods. *Int J Res Phar BiomedSci*. 2011;2:680–686.

Vitzthum OG, Werkhoff P, Hubert P. New volatile constituents of black tea aroma. *J Agric Food Chem*. 1975; 23: 999-1003.

Wang JJ, Liao CC, Hu WP, Shen HC. *Synthesis of 2-(4-aminophenyl) benzothiazole derivatives and use thereof*. US0215154; 2012.

Wang L, Doherty G, Wang X, et al., *Preparation of substituted benzothiazoles as apoptosis-inducing agents for the treatment of cancer, immune and autoimmune diseases.* WO055895; 2013.

Wei G, Yang B, He Q. Ubiquitin dependent degradation of CDK2 drives the therapeutic differentiation of AML by targeting PRDX2. *Blood*. 2018;131:2698-2711.

White EH, McCapra F, F. Field GF. The structure and synthesis of firefly Luciferin. *J Am Chem Soc*. 1963; 85: 337-343.

Xue W, He Y, Chen Q, et al., *Preparation of N-(2-(substituted benzothiazole-2-carbamoyl)-phenyl)- benzamide derivatives as antiphytoviral and antitumor agents*. CN102391207; 2012.

Yaginuma S, T. Asahi, M. Takada, M. Hayashi, K. Mizuno, Japanese Patent 1181793-A; 1989.

Ying M, X. Shao, H. Jing, Y. Liu, X. Qi, J. Cao, Y. Chen, S. Xiang, H. Song, R. Hu,Malik JK, Nanjwade BK, Noolvi MN, et al., Novel diaryl substituted imidazo (2,1-b)benzothiazole derivatives, process for their preparation, and their pharmacological evaluation. INMU02510; 2011.

Hu Y, MacMillan JB. Erythrazoles A–B, Cytotoxic benzothiazoles from a marine-derived *Erythrobacter sp.Org Lett.* 2011;13:6580-6583.

Zhang C, Xu D, Wang J, Kang C. Efficient synthesis and biological activity of novel indole derivatives as VEGFR-2 tyrosine kinase inhibitors. *Russ J Gen Chem.* 2017;87(12):3006–3016.

Zhang L, Fan J, Vu K, Hong K, Le Brazidec JY, Shi J et al., 7′-substituted benzothiazolothio- and pyridinothiazolothio-purines as potent heat shock protein 90 inhibitors. *J Med Chem* 2006;49:5352–5362.

Zhu ML, Wang CY, Xu CM, Bi WP, ZHou XY. Evaluation of 6-chloro-N-(3,4-disubstituted-1,3-thiazol-2(3*H*)-ylidene)-1,3-benzothiazol-2-amine using drug design concept for their targeted activity against colon cancer cell lines HCT-116, HCT15, and HT29. *Med Sci Monit.* 2017;23:1146-1155.

Zou Q, Liu Y, Zhengwei Y, et al., *Preparation of heterocycles as protein kinase inhibitors.* WO127345; 2013.

INDEX

#

2-anilino benzothiazole derivatives, 82, 83, 84, 89

A

absorption spectroscopy, 164
access, 44, 57, 58, 61, 64, 66
acetaminophen, 30, 48
acetic acid, 2, 6, 78, 80, 129
acetone, 120, 121
acetonitrile, 111, 171
acetylation, 153
acid, vii, viii, 1, 4, 9, 11, 14, 29, 34, 58, 65, 68, 71, 72, 75, 79, 80, 81, 90, 92, 103, 118, 119, 152, 153, 167, 179, 181, 182
acidic, 24, 45, 101
active compound, 83, 90
active site, 147
acyl, 7, 76, 86, 88, 89, 90, 93, 119, 167
additives, 15, 111
adenocarcinoma, 87, 165, 170

agonist, 21, 126, 183, 189
agriculture, 135
alanine, 30, 141
alanine aminotransferase, 30
aldehydes, vii, 1, 7, 8, 9, 10, 27, 54, 133, 171
alkaline phosphatase, 118, 162
alkaloids, 2, 5, 31, 133, 178, 181
allergic reaction, 29
amine, 66, 73, 81, 82, 89, 90, 103, 147, 149, 165, 171, 191
amino, 4, 21, 27, 41, 42, 55, 73, 74, 76, 84, 85, 91, 94, 104, 140, 141, 148, 150, 152, 154, 155, 156, 157, 165, 173, 174, 178, 179, 183, 185
amino acid, 84, 141, 148, 150, 178, 183
ammonium, 10, 35
amylase, 21, 31
amyotrophic lateral sclerosis, viii, 2, 17, 31, 125, 186
analgesic, vii, viii, 1, 25, 112, 121
analgesic agent, 25
angiogenesis, 55, 144, 146, 150
anhydrase, 61, 122

aniline, 11, 35, 61, 80, 174
antibiotic, 6, 188
anticancer, v, vii, viii, ix, 1, 16, 18, 19, 37, 38, 40, 49, 54, 55, 57, 58, 59, 60, 63, 68, 71, 72, 76, 82, 83, 84, 85, 86, 89, 91, 92, 93, 94, 112, 117, 118, 121, 136, 137, 138, 139, 142, 143, 144, 147, 149, 150, 151, 154, 156, 158, 161, 164, 165, 166, 167, 168, 169, 170, 172, 173, 174, 175, 176, 177, 178, 179, 180, 182, 183, 184, 185, 187, 190
anticancer activity, vii, viii, 37, 38, 63, 72, 76, 82, 83, 84, 86, 90, 92, 94, 143, 144, 147, 156, 165, 167, 170, 172, 173, 174, 175, 178, 179, 183, 184
anticancer drug, 143, 154, 164, 169, 178
anticonvulsant, v, vii, viii, 1, 16, 21, 41, 42, 71, 72, 76, 77, 79, 80, 89, 91, 92, 93, 112, 121, 135, 184
antidepressant, 121
antidiabetic, vii, viii, 1, 16, 21, 42, 74, 92, 121, 126
antiepileptic drugs, 93
anti-inflammatory, vii, viii, 1, 16, 24, 25, 44, 45, 67, 70, 73, 91, 112, 121, 125
anti-inflammatory agents, 25, 45, 67
antimalarial, 22, 75, 92, 112
antimicrobial, vii, viii, 1, 16, 23, 24, 41, 43, 44, 45, 56, 61, 64, 73, 92, 112, 121, 154, 185, 188
antioxidant, 75, 92, 96, 119, 121, 165, 170, 171, 187, 190
antipsychotic, 25, 46
antitumor, viii, 18, 25, 37, 38, 39, 71, 72, 76, 83, 86, 93, 94, 124, 133, 136, 137, 138, 139, 140, 142, 147, 155, 163, 164, 169, 176, 178, 179, 180, 181, 182, 183, 185, 188, 189, 191
antitumor agent, viii, 18, 19, 37, 71, 72, 76, 83, 86, 94, 136, 137, 138, 139, 140, 155, 163, 176, 178, 180, 185, 189, 191

apoptosis, 50, 55, 56, 57, 58, 59, 60, 62, 66, 86, 94, 138, 146, 153, 157, 159, 163, 164, 167, 169, 170, 177, 180, 183, 185, 186, 190
arrest, 153, 157, 158, 164, 182
atoms, viii, ix, 71, 76, 117, 120, 121
autoimmune diseases, 138, 190

B

barbituric, 79, 80, 81, 90, 92
base, 112, 120, 152, 159, 171, 187
benzene, vii, viii, ix, 1, 2, 13, 71, 72, 73, 75, 100, 117, 120, 121, 140
benzodiazepine, 49, 137, 183
benzothiazole, v, vii, viii, ix, 1, 2, 3, 4, 5, 6, 7, 15, 16, 17, 18, 19, 20, 21, 22, 23, 24, 25, 26, 27, 28, 29, 31, 33, 37, 38, 39, 40, 41, 42, 43, 44, 45, 46, 48, 50, 55, 56, 57, 65, 66, 71, 72, 73, 74, 75, 76, 77, 78, 79, 80, 81, 82, 83, 84, 85, 86, 88, 89, 90, 91, 92, 93, 94, 95, 99, 100, 101, 102, 103, 105, 106, 107, 108, 109, 110, 111, 112, 113, 117, 118, 120, 121, 122, 127, 129, 133, 134, 135, 136, 137, 138, 139, 140, 142, 143, 145, 146, 147, 148, 149, 150, 151, 152, 153, 154, 155, 156, 157, 158, 159, 160, 161, 162, 163, 164, 165, 166, 167, 168, 169, 170, 171, 172, 173, 174, 175, 176, 177, 178, 179, 180, 181, 182, 183, 184, 185, 186, 187, 188, 190, 191
binding energy, 150
bioaccumulation, 31
bioassay, 165
bioavailability, 140, 141, 144, 162
biocatalysts, 51
biological activities, 2, 5, 16, 72, 77, 84, 112, 174, 180
biological activity, 26, 42, 45, 191
bioluminescence, 6
biomolecules, 120

Index

biosynthesis, 2, 5, 56
bladder cancer, 30, 82
blood vessels, 146
bonding, viii, 100, 108, 149
brain, 87, 112, 184
breast cancer, 62, 66, 125, 140, 144, 162, 165, 169, 171, 179, 189
breast carcinoma, 180
bromine, 13, 80, 83

C

C_7H_5NS, vii, 1, 120, 121
cancer, viii, 18, 30, 39, 49, 51, 60, 71, 72, 82, 83, 85, 87, 90, 94, 136, 137, 138, 139, 141, 143, 144, 147, 148, 150, 151, 152, 153, 156, 158, 159, 160, 162, 163, 164, 166, 167, 168, 169, 170, 171, 175, 176, 178, 180, 181, 182, 183, 185, 186, 188, 190
cancer cells, 87, 90, 94, 141, 150, 160, 162, 168, 169, 170, 182
cancer death, 82
cancer therapy, 82, 163, 186
cancerous cells, 173
carbohydrate, 31, 68
carbon, 16, 66, 68, 120
carboxylic acid, vii, 1, 7, 94, 134, 171
carboxylic acids, vii, 1, 7
carcinoma, 29, 56, 140, 144
catalyst, 9, 10, 11, 12, 15, 62, 157, 169
catalytic system, 61, 162
cell cycle, 153, 157, 159, 163, 164
cell death, 63, 87, 141, 157, 161, 162, 171, 182
cell differentiation, 163
cell division, 163
cell line, 7, 18, 39, 58, 63, 82, 83, 85, 87, 90, 133, 136, 143, 144, 146, 147, 148, 152, 153, 156, 158, 159, 162, 163, 164, 165, 167, 168, 169, 170, 171, 175, 177, 182, 186, 189, 191
cell signaling, 151
cervical cancer, 147, 157, 159, 165, 167, 171, 173, 186
chemical, vii, 1, 17, 27, 30, 49, 51, 68, 101, 121, 147, 173, 175
chemical properties, 17
chemical structures, 147, 175
chemicals, 30, 47, 48
chemotherapeutic agent, ix, 18, 118, 136, 141, 142, 143, 157, 162, 176, 184
chemotherapy, 49, 154, 169, 175
chlorobenzene, 13
chloroform, 83, 108
chronic obstructive pulmonary disease, 17, 183
cleavage, 58, 63, 64, 66, 68
clinical trials, ix, 17, 18, 118, 135, 136, 141, 152, 156
colon, 82, 136, 140, 144, 147, 148, 156, 162, 170, 181, 191
colon cancer, 162, 170, 191
colorectal adenocarcinoma, 87
colorectal cancer, 37, 147, 168, 177
compounds, 17, 20, 22, 27, 28, 33, 35, 37, 38, 43, 49, 54, 60, 67, 73, 77, 80, 82, 85, 86, 91, 93, 104, 108, 110, 111, 112, 114, 119, 133, 136, 137, 138, 139, 144, 146, 148, 150, 152, 154, 155, 156, 157, 158, 159, 161, 162, 164, 165, 167, 168, 169, 170, 171, 172, 173, 176, 178, 181, 182, 183, 184, 185, 186, 187
condensation, 7, 8, 10, 133
conjugation, 20, 30, 141
constituents, 27, 33, 188, 190
crystal structure, 40, 42, 104, 179
cytochrome, 19, 138, 152, 178, 186, 187
cytochrome p450, 178
cytometry, 169
cytotoxic agents, 55

cytotoxicity, 7, 29, 47, 54, 55, 62, 63, 87, 153, 154, 156, 157, 159, 167, 168, 171, 173, 174, 187

D

degradation, 152, 167, 191
derivatives, vii, viii, ix, 1, 2, 5, 6, 15, 17, 18, 19, 20, 21, 22, 23, 24, 25, 26, 28, 31, 37, 38, 40, 41, 42, 43, 44, 45, 46, 55, 56, 61, 62, 63, 70, 71, 72, 73, 74, 75, 76, 77, 78, 79, 80, 81, 83, 84, 85, 86, 88, 89, 91, 92, 93, 94, 95, 100, 108, 111, 113, 118, 121, 133, 134, 135, 136, 137, 138, 139, 140, 141, 142, 143, 145, 146, 147, 148, 149, 150, 151, 153, 154, 155, 156, 157, 158, 160, 161, 162, 163, 164, 165, 166, 167, 168, 169, 172, 173, 175, 176, 177, 179, 182, 183, 184, 185, 186, 187, 188, 189, 190, 191
dermatitis, 29
detection, 112, 113
digestive enzymes, 31
dimethylformamide, 85
diseases, viii, 71, 72, 82, 91, 137, 177, 189
diuretic, 26, 46, 121, 122, 135
diversity, 2, 67
DNA, viii, 19, 29, 36, 37, 40, 50, 55, 56, 57, 58, 59, 60, 63, 71, 72, 87, 141, 154, 156, 157, 159, 164, 167, 179, 182, 183, 185, 186, 187
DNA damage, 159, 167, 182
DNA strand breaks, 141
down-regulation, 164
drinking water, 48
drug addict, 26
drug addiction, 26
drug design, ix, 20, 39, 56, 118, 121, 191
drug discovery, ix, 16, 18, 19, 36, 112, 118, 119, 121
drug interaction, 143

drugs, viii, ix, 2, 24, 31, 74, 77, 80, 90, 118, 120, 121, 135, 144
dyes, 2, 17, 27, 31, 111, 136, 182

E

electron, 13, 26, 101, 103, 108, 110, 112, 147, 157
emission, 101, 103, 105, 108, 110, 111, 112, 113
environment, viii, 2, 51, 100, 102
enzyme, 8, 19, 36, 40, 70, 141, 157
enzyme inhibitors, 36
enzymes, 8, 30, 48, 51, 95, 144, 152, 153, 154, 178
epilepsy, viii, 71, 72, 76, 91
eukaryotic, 38, 154, 156, 187

F

family members, 161
fatty acids, 12
firefly luciferin, 2, 6, 33
fluorescence, viii, 100, 101, 102, 103, 105, 108, 110, 111, 112, 113
fluorine, 73, 140, 153
formation, 13, 39, 54, 62, 63, 140, 152, 167, 169
free radicals, 171
functionalization, 54, 63, 66

G

geometry, 101, 110, 111
glioblastoma, 125, 148
glucose, 10, 74
glucose oxidase, 10
glutathione, 118, 170

growth, 26, 50, 75, 82, 93, 119, 140, 143, 146, 147, 148, 150, 151, 165, 167, 168, 171
growth factor, 82, 93, 119, 146, 150
growth signal, 143

H

halogen, 155, 173, 174
heat shock protein, 160, 161, 188, 191
heterocyclic, vii, ix, 1, 32, 37, 43, 49, 50, 67, 68, 72, 100, 112, 114, 117, 118, 119, 120, 170, 173, 179, 182, 184, 187
histone, 153, 182
histone deacetylase, 153, 182
human, viii, 18, 29, 37, 62, 66, 72, 82, 87, 93, 94, 103, 136, 140, 143, 148, 152, 153, 156, 157, 158, 162, 163, 164, 165, 167, 169, 170, 171, 179, 180, 181, 186, 187, 190
hybrid, 20, 163, 170
hybridization, 54, 169
hydrazine, 75, 77, 78, 89, 169, 173
hydrazones, 55, 75, 76, 77, 86, 88, 89, 90, 92, 93, 94
hydrocarbons, 27
hydrogen, 86, 101, 104, 108, 112, 147, 149, 157
hydrogen bonds, 147
hydroxyl, 108, 147, 157, 171, 173, 174
hydroxyl groups, 157, 174
hypothesis, 119

I

immunoglobulin, 146
immunomodulatory, 128, 133
in vitro, 18, 22, 28, 36, 37, 38, 43, 47, 54, 55, 56, 85, 96, 128, 140, 146, 148, 163, 164, 165, 167, 169, 171, 181, 182, 185, 187, 190

in vivo, 28, 36, 41, 46, 47, 128, 140, 149, 152, 153, 185, 187, 189
India, 1, 48, 52, 53, 66, 69, 71, 95, 99, 117
induction, 141, 153, 157, 167
industrial applications., 2, 31
industrial chemicals, 26, 27
industries, 17, 51
industry, 17, 26, 27, 30, 48, 50, 51, 112, 135
inhibition, 19, 37, 38, 40, 55, 56, 57, 58, 60, 62, 70, 85, 87, 137, 143, 146, 147, 148, 150, 152, 153, 154, 157, 162, 169, 170, 171, 174, 183, 184, 187, 189
inhibitor, 6, 20, 122, 126, 135, 137, 143, 144, 150, 151, 154, 161, 164, 177, 179, 180, 185, 190

K

ketones, 27, 28, 58, 66, 78
kidney, 147, 148, 173

L

lipase catalysed transesterification, 50
liquid crystals, 17
liquid phase, 68
luciferin, 2, 6, 33, 133
lung cancer, 7, 30, 58, 63, 83, 90, 119, 133
lymph, 30, 47
lymph node, 30, 47
lymphangiogenesis, 176
lymphocytes, 190
lymphoma, 118
lysine, 141, 152

M

matrix metalloproteinase, 138, 183
metabolism, 48, 94, 140, 151, 152, 154
metabolized, 94

metabolizing, 30, 48
metastasis, 143, 146, 150
methyl group, 140, 153
mice, 30, 77, 80, 91, 159, 162
mitosis, 157, 158, 160
models, 30, 159, 168, 180
molecular dynamics, 55
molecular structure, 93, 101
molecules, vii, viii, ix, 2, 36, 67, 68, 71, 72, 76, 100, 101, 103, 109, 113, 118, 119, 136, 140, 146, 149, 150, 163, 170
multicellular organisms, 154
muscle relaxant, 25, 112

N

natural compound, 2, 5, 6, 31, 112
neurodegeneration, viii, 71, 72
neurodegenerative disorders, 112
neurological disease, 76
neurotoxicity, 29, 77, 80, 90, 91
neurotransmission, 84
nitrogen, ix, 75, 100, 112, 117, 120, 121, 149
nonlinear optics, 2, 17, 31
non-steroidal anti-inflammatory drugs, 24
nucleus, viii, ix, 1, 3, 23, 100, 111, 114, 118, 121, 122, 134, 142, 147, 173
nutrients, 146, 150

O

oncogenes, 153, 160
optimization, 61, 111, 140, 141, 190
organic compounds, 103
ovarian cancer, 161, 165, 171
oxidation, 39
oxidation products, 39
oxidative agents, 7
oxide nanoparticles, 97
oximes, 68

oxygen, 10, 119, 146, 149, 161

P

pancreatic cancer, 170, 190
patents, 54, 137, 140, 174
pathway, 57, 62, 66, 87, 151, 167, 184, 189
pharmaceutical, 2, 32, 51, 100, 112, 137, 138, 139, 152, 177, 179, 182, 187
pharmaceutical applications, 2, 112
pharmacokinetics, 57, 150
pharmacology, 47, 91
phase-I, 17, 18, 30, 80
phosphate, 9, 34, 113
phosphorylation, 144
photo-excitation, 101
photo-physical, vii, viii, 99, 100, 111
photosensitizers, 27
physical properties, vii, viii, 99, 111, 121
physicochemical properties, 144
polymerization, 38, 50, 54, 57, 58, 59, 60, 63, 65, 66, 157, 158, 159, 184
positron emission tomography, 177
pramipexole, viii, 1, 2, 17, 31, 124, 190
preparation, iv, 7, 28, 33, 137, 138, 178, 183, 185, 189, 191
prodrugs, 49, 62, 152, 176, 178, 183
proliferation, 19, 37, 82, 153, 159, 160, 170, 184
prostate cancer, 60, 162
protection, 27, 77, 80, 90, 178
protein-protein interactions, 188
proteins, 144, 153, 159, 160, 178
pyrimidine, 24, 44, 73, 79, 85, 90, 91, 137, 163, 177, 179
pyrimido, 76, 84, 85, 86, 90, 92

Q

quadruplex DNA, 62
quantum yields, ix, 100, 110, 111

Index

quercetin, 144

R

radicals, 75, 171
reactions, 7, 8, 10, 12, 13, 29, 68, 101, 179, 184
receptor, 21, 26, 82, 93, 119, 125, 126, 146, 147, 150, 164, 177, 181, 183
resistance, 82, 150, 184, 188
restless legs syndrome, 124
rheumatoid arthritis, 123, 135
riluzole, viii, ix, 1, 2, 17, 31, 41, 71, 72, 84, 91, 118, 125, 135, 186
room temperature, 14, 121
rubber, 27, 29, 30, 47, 48, 135

S

selectivity, 24, 49, 87, 112, 140, 141, 173, 176
semicarbazones, 76, 77, 78, 79, 89, 92
sensitivity, 37, 47, 83, 172
shock, 119, 178, 189
showing, 90, 102, 112
side effects, 49, 77, 143, 156, 159
signal transduction, 144, 146
signaling pathway, 151
skeleton, ix, 86, 118, 121
sodium, 13, 14, 15, 73, 78
solar cells, 17
solar collectors, 111
solid phase, 67, 68
solid state, ix, 100, 104, 106, 108, 110
solubility, 31, 103, 141, 152, 162
solution, ix, 68, 100, 110, 113
solution phase method, 68
solvent, viii, 8, 9, 10, 11, 12, 14, 34, 79, 80, 85, 100, 106, 108, 110, 157
solvent dissociation, 109
solvent free, 9, 12

solvents, 108, 110, 112
species, 96, 101, 112, 119, 141
spectrophotometric method, 190
sponge, 6, 33, 127, 128, 131, 133, 181, 189
structural modifications, 39, 49
structure, vii, ix, 3, 20, 32, 38, 86, 94, 100, 104, 109, 112, 118, 119, 120, 121, 122, 123, 124, 125, 126, 127, 128, 129, 130, 131, 132, 140, 144, 145, 146, 148, 149, 151, 153, 154, 155, 156, 158, 160, 161, 162, 163, 164, 166, 168, 169, 172, 175, 176, 180, 181, 191
substitution, ix, 26, 87, 104, 110, 118, 121, 147, 151, 153, 154, 156, 157, 161, 169, 170, 174
sulfonamide, 24, 44, 61, 73, 184
sulfur, ix, 6, 40, 75, 100, 117, 120, 121, 150, 161, 173
synthesis, v, vii, ix, 1, 2, 6, 7, 8, 9, 10, 11, 12, 13, 14, 15, 16, 27, 32, 33, 34, 35, 36, 37, 38, 39, 40, 41, 42, 43, 44, 45, 46, 49, 50, 54, 55, 56, 57, 58, 59, 60, 61, 62, 63, 64, 65, 66, 67, 68, 69, 70, 78, 80, 81, 83, 84, 85, 86, 87, 88, 91, 92, 93, 94, 95, 97, 100, 101, 111, 112, 118, 121, 133, 134, 138, 143, 144, 146, 147, 148, 152, 154, 161, 168, 169, 171, 176, 177, 178, 179, 180, 181, 182, 183, 184, 185, 186, 187, 188, 189, 190, 191

T

target, 19, 49, 60, 68, 74, 119, 143, 144, 146, 147, 150, 151, 153, 154, 157, 162, 163, 164, 176, 180
therapeutic agents, viii, 1, 17, 121, 122, 138, 175, 189
therapeutic targets, 168
therapeutic use, 138, 187
therapeutics, 51, 52, 68
therapy, 60, 161, 186

treatment, 26, 31, 82, 91, 97, 126, 135, 137, 138, 139, 143, 144, 150, 152, 167, 177, 178, 179, 181, 182, 183, 185, 189, 190
trial, viii, 72, 125
tuberculosis, 19, 20, 21, 40, 70
tumor, 82, 112, 135, 141, 146, 150, 151, 152, 153, 159, 162, 165, 167, 168, 171, 182
tumor cells, 82, 146, 153
tumor growth, 82, 146, 159
tumorigenesis, 144, 150, 180
tumour growth, 162
tyrosine, 82, 144, 146, 147, 150, 173, 178, 181, 188, 191

V

various aldehydes, 9
vascular endothelial growth factor, 146
viscosity, viii, 100, 111
vulcanization, 2, 17, 27, 30, 31, 112, 135

W

water, 10, 27, 31, 35, 68, 97, 103, 111, 120, 127, 128, 133, 141, 152, 181
wavelengths, 111
wood, 6, 27, 33, 127
workers, 29, 30, 103, 133, 143, 147, 150, 153, 163, 165, 168
worldwide, 18, 76, 82
wound healing, 150

X

xanthones, 96

Y

yield, 9, 10, 11, 12, 16, 80, 83, 85, 103, 105, 108, 110, 111, 133, 174

Z

zirconium, 9, 34